Springer Series on
SIGNALS AND COMMUNICATION TECHNOLOGY

Signals and Communication Technology

Passive Eye Monitoring
Algorithms, Applications and Experiments
R.I. Hammoud (Ed.) ISBN 978-3-540-75411-4

Multimodal User Interfaces
From Signals to Interaction
D. Tzovaras ISBN 978-3-540-78344-2

Human Factors and Voice Interactive Systems
D. Gardner-Bonneau, H.E. Blanchard (Eds.)
ISBN: 978-0-387-25482-1

Wireless Communications
2007 CNIT Thyrrenian Symposium
S. Pupolin (Ed.) ISBN: 978-0-387-73824-6

Satellite Communications and Navigation Systems
E. Del Re, M Ruggieri (Eds.)
ISBN: 978-0-387-47522-6

Digital Signal Processing
An Experimental Approach
S. Engelberg ISBN: 978-1-84800-118-3

Digital Video and Audio Broadcasting Technology
A Practical Engineering Guide
W. Fischer ISBN: 978-3-540-76357-4

Three-Dimensional Television
Capture, Transmission, Display
H.M. Ozaktas, L. Onural (Eds.)
ISBN 978-3-540-72531-2

Foundations and Applications of Sensor Management
A.O. Hero, D. Castañón, D. Cochran,
K. Kastella (Eds.) ISBN 978-0-387-27892-6

Digital Signal Processing with Field Programmable Gate Arrays
U. Meyer-Baese ISBN 978-3-540-72612-8

Adaptive Nonlinear System Identification
The Volterra and Wiener Model Approaches
T. Ogunfunmi ISBN 978-0-387-26328-1

Continuous-Time Systems
Y.S. Shmaliy ISBN 978-1-4020-6271-1

Blind Speech Separation
S. Makino, T.-W. Lee, H. Sawada (Eds.)
ISBN 978-1-4020-6478-4

Cognitive Radio, Software Defined Radio, and Adaptive Wireless Systems
H. Arslan (Ed.) ISBN 978-1-4020-5541-6

Wireless Network Security
Y. Xiao, D.-Z. Du, X. Shen
ISBN 978-0-387-28040-0

Terrestrial Trunked Radio – TETRA
A Global Security Tool
P. Stavroulakis ISBN 978-3-540-71190-2

Multirate Statistical Signal Processing
O.S. Jahromi ISBN 978-1-4020-5316-0

Wireless Ad Hoc and Sensor Networks
A Cross-Layer Design Perspective
R. Jurdak ISBN 978-0-387-39022-2

Positive Trigonometric Polynomials and Signal Processing Applications
B. Dumitrescu ISBN 978-1-4020-5124-1

Face Biometrics for Personal Identification
Multi-Sensory Multi-Modal Systems
R.I. Hammoud, B.R. Abidi, M.A. Abidi (Eds.)
ISBN 978-3-540-49344-0

Cryptographic Algorithms on Reconfigurable Hardware
F. Rodríguez-Henríquez
ISBN 978-0-387-33883-5

Ad-Hoc Networking Towards Seamless Communications
L. Gavrilovska ISBN 978-1-4020-5065-7

Multimedia Database Retrieval
A Human-Centered Approach
P. Muneesawang, L. Guan
ISBN 978-0-387-25627-6

Broadband Fixed Wireless Access
A System Perspective
M. Engels; F. Petre
ISBN 978-0-387-33956-6

Acoustic MIMO Signal Processing
Y. Huang, J. Benesty, J. Chen
ISBN 978-3-540-37630-9

Algorithmic Information Theory
Mathematics of Digital Information Processing
P. Seibt ISBN 978-3-540-33218-3

Continuous-Time Signals
Y.S. Shmaliy ISBN 978-1-4020-4817-3

Interactive Video
Algorithms and Technologies
R.I. Hammoud (Ed.) ISBN 978-3-540-33214-5

Handover in DVB-H
Investigation and Analysis
X. Yang ISBN 978-3-540-78629-0

Xiaodong Yang

Handover in DVB-H

Investigation and Analysis

Springer

Dr. Xiaodong Yang
Gentzgasse 107-3
1180 WIEN
Austria
xyang@informatik.uni-goettingen.de
eepgxxy@yahoo.co.uk

ISBN 978-3-642-09738-6 e-ISBN 978-3-540-78630-6

DOI 10.1007/978-3-540 78630-6

Springer Series on Signals and Communication Technology ISSN 1860-4862

Softcover reprint of the hardcover 1st edition 2008

Cover design: WMXDesign GmbH, Heidelberg

Printed on acid-free paper

9 8 7 6 5 4 3 2 1

springer.com

I would like to dedicate this book to the hardworking and still innovating DVB-H community.

Preface

"European industry has already developed successful standards in the past, and I am very confident that on the basis of DVB-H, Mobile TV services can develop the economies of scale they need for take-up across Europe and around the world," With these words of EU's Telecom Commissioner Viviane Reding, DVB-H is destined to be a dominating mobile TV technology in Europe and even in the world.

I was first getting in touch with the DVB technology when I was doing my PhD research in Brunel University in UK in 2002. At that time DVB-T was already a mature and widely used digital broadcast technology and anyone could easily buy a DVB-T receiver in the market to try the digital broadcast signals that have been already broadcasted in UK since 1998. Then the DVB technology world changed dramatically. As a more flexible and robust terrestrial broadcast system targeting handsets, DVB-H was developed based on DVB-T. In 2003 the DVB-H community were continuously working to finalize the standard. Finally in November 2004 DVB-H was adopted as an ETSI standard EN 302 304. I was lucky to see all these changes when I was doing my PhD research in DVB technology. And I was very proud to be involved in the different DVB-H research projects since the beginning of the DVB-H standard development stage. I was also lucky enough that I am one of the first persons who finished PhD degree by focusing on DVB-H research. The more I was involved in the DVB-H research, the more I realized that there was a shortage of books which can systematically introduce the DVB-H technology to researchers, engineers and all those who are interested in this technology. Therefore I decided to write a comprehensive book about DVB-H by focusing on the DVB-H handover technology. No books about handover technology in DVB-H was available up to writing of this book. As one of the main persons in the world who are doing the handover technology research in DVB-H, I attempted to fill this gap in the literature.

DVB-H is the broadcast technology that broadcasts IP data packets to the handheld devices. Due to its broadcast nature DVB-H can support large scale consumption of Mobile TV that the telecommunication technology such as 3G can no longer accommodate. In order to have a large mobile reception coverage more transmitters and repeaters (gap fillers) are needed for DVB-H than for DVB-T. In this case, multi-frequency network structure will be one of the main network topology for DVB-H in the future. As a result, just like traditional telecommunication technologies, handover in DVB-H is also necessary when the users move from one DVB-H cell to another. As a novel mobile broadcast technology, DVB-H blurred the traditional border between telecommunications domain and the broadcast domain. In this background, a novel network structure that combined telecommunications and broadcast technology was created, which is called converged network. Handover in such converged networks became also a hot research topic.

This book will focus on the handover technology in DVB-H and in the converged networks between DVB-H and UMTS. As it also gives much introduction and analysis for DVB-H handover related information, for example ESG, it is a must-have book for any person who is not only interested in the DVB-H handover but also in DVB technology in general. Each chapter of the book is complete and independent which can be read independently by those who are interested in only some particular topics. By reading the whole book, the readers will see a complete picture of DVB-H technology. At the end of each chapter, there are some questions which are mostly asked by others to me when I am doing the research and I believe they will probably also come to the minds of the readers when the readers read the chapter. And at the end of the book, there are solutions to the questions raised in each chapter. This book can be used by broadcast and telecommunications researchers, engineers, academics, regulatory bodies and business managers as a reference book, or by university students as a text book or a reference book.

The chapter structure of this book is as follows:

Chapter 1 introduces the DVB-H technology, its evolution and technical features, its network components and network structure.

Chapter 2 presents the motivation of the handover research in DVB-H and the approaches used to address the handover problems in the DVB-H research.

Chapter 3 provides a comprehensive survey of the research that has been conducted on the handover issues in DVB-H networks.

Chapter 4 presents a comprehensive introduction of the signalling information in DVB-H and pointed out which signalling information can be utilized for the DVB-H handover.

Chapter 5 is a chapter focusing on the Electronic Service Guide in DVB-H. It also points out how the Electronic Service Guide can be used in the handover in DVB-H.

Chapter 6 presents different handover algorithms for dedicated DVB-H networks. General introduction and analysis for the handover in dedicated DVB-H networks are given.

Chapter 7 focuses on the handover algorithm based on post processing of SNR values for a dedicated DVB-H network.

Chapter 8 presents the repeater aided handover algorithm for a dedicated DVB-H network.

Chapter 9 provides the soft handover probability calculation of the repeater aided handover algorithm.

Chapter 10 introduces the handover in the converged networks. As an example, the handover algorithm between DVB-H and UMTS in the converged network is presented. The stochastic trees model for such handover is used and analyzed.

Chapter 11 introduces the handover in the hybrid broadcast networks. The vertical handovers between different broadcast technologies such as between DVB-H and DMB are presented and analyzed.

Chapter 12 concludes the book by giving a comparison of the different handover algorithms in the dedicated DVB-H networks. It also presents some of the future research topics of DVB-H and DVB-H handover technology.

I believe this book will help raise new research problems and bring new solutions in the DVB or other multimedia communication technologies. Any comments to improve the book will be highly appreciated.

The last but not the least I would like to thank all those people who have helped and advised me in my research in DVB-H. Special thanks are given to EU project IST INSTINCT and IST MING-T which have given great impulse to my research.

Vienna, Austria

Xiaodong Yang
January 2008

Contents

1 Introduction

The communication world is evolving quickly with new technologies being continuously developed. In the telecommunication world, UMTS is a 3G standard for end-to-end mobile systems utilizing Wideband Code Division Multiple Access (WCDMA) access technology. In the broadcast world, Digital Video Broadcasting - Terrestrial (DVB-T) is a leading international standard established for the move from analogue to digital broadcasting basically via terrestrial networks. Within DVB-T, it is possible to carry defined data containers in addition to the audio and video in the Motion Picture Experts Group Technical Standard-2 (MPEG-2) transport stream. These data containers can be used to realize new data services or to carry IP datagrams. Thus it enables large amounts of data to be delivered to a lot of users in both fixed and mobile environments.

A terrestrial broadcasting channel differs from a satellite transmission link or a cable channel in that it is prone to multipath propagation. Reflections of the transmitted signal from obstacles such as buildings or mountains are superimposed asynchronously on the directly received signal. The OFDM modulation scheme was introduced to DVB standards to utilize the multipath effects to generate stronger signals instead of generating inferences. The performance of the OFDM scheme in DVB-T is analysed in [107].

DVB-H (formerly "DVB-X" [7]) together with DVB-T were developed by the DVB Project [51]. DVB-H targets handheld mobile terminals [108, 9]. As the multi-frequency cellular structured DVB-H network is a typical network structure, handover becomes a critical issue for DVB-H. Handover in DVB-H refers to the synchronization of the transport stream and the frequency when the terminal moves from one cell to another.

1.1 Telecommunication and Broadcast

In recent years, both the telecommunications and the broadcast industries have made the analogue-to-digital transition with Global System for Mobile

Telecommunications/Universal Third Generation (GSM/3G) [43, 44] and Digital Audio Broadcasting/Digital Video Broadcasting (DAB/DVB)[50, 51] respectively. 3G telecommunication networks like Universal Mobile Telecommunications System (UMTS) [52] have been able to provide much higher data rates than either the GSM or General Packet Radio Service (GPRS)networks. Multimedia Broadcast Multicast Service (MBMS) [65] is rolling out to provide the same light-duty service to multiple users simultaneously.

On the other hand, the audio/video encoding technology is developing quickly. Thus now the users can have a High Definition Television (HDTV) viewing experience by watching streaming service on a mobile phone. In this case, the streaming service is delivered simply using H.264 video encoding and MPEG-4 High Efficiency Advanced Audio Coding (HE AAC) technology with 10 frames per second (fps) rate, 200K bite rate per second (bps) and Quarter Video Graphics Array (QVGA) video format. And the radio bear technology behind this is simply UMTS.

However, UMTS is designed to provide "point to point" unicast service to users. This means, when there are a lot of users trying to access the same streaming service, the UMTS bandwidth will not be able to accommodate this and the network congestion will happen. Besides, an advanced viewing experience using 15-25 fps, 384K bps is nearly impossible for simultaneous multi-user reception in UMTS. Thus full heavy-duty video streaming and download are still not viable in 3G technology for a variety of reasons including the cost to the users.

Because of the demand for a better viewing experience, people began to seek help from the tradition broadcast technology. Broadcast technology is a "one to many" technology. Digital Video Broadcasting-Terrestrial (DVB-T) [45] and Digital Video Broadcasting-Handheld (DVB-H) [1, 161] are developed to bring low cost multimedia services to numerous users at the same time without inflicting additional burden to the transmission networks.

DVB-H is targeting handheld receivers since the beginning of its development. Depending on the underlying modulation technology, it can be used to easily transmit 10-20 high definition video channels with 384 Kbps, 25 fps and QVGA video format within a 8MHz spectrum bandwidth.

While both DVB-H and 3G technologies are targeting handheld receivers such as mobile phones, the traditional border between broadcast and telecommunication become blurred. On the other hand, such situation results in a novel network concept - converged networks, which brings the broadcast and telecommunication technology together and makes them converge. In the converged networks, the downlink audio/video can be delivered using either the broadcast networks (e.g. DVB-H) or the telecommunications networks (e.g. 3G), while the uplink will utilize the telecommunications networks (e.g. 3G). The converged network can thus deliver low cost multimedia data services to multi-users like the broadcast networks while at the same time providing interactivity like the telecommunication networks.

1.2 Handover in DVB-H

Handover is always an important issue in telecommunications networks [41]. DVB-H is developed from DVB-T with added features suitable for limited battery powered handheld terminals in mobile environments. The modulation scheme in DVB-H is Orthogonal Frequency Division Multiplexing (OFDM). There are three modes being used: 2K, 4K and 8K which stands for different number of carriers. The 2K mode employs 1705 individual carriers, the 4K mode employs 3409 carriers and the 8K mode employs 6817 carriers [31, 127]. DVB-H cell size is variable depending on the OFDM modulation modes. It is up to 17km for the 2K mode and up to 67km for the 8K mode [128]. Although single frequency netwok (SFN) will be a main network structure for DVB-H at the first deployment phase, multi-frequency networks (MFN) will also be a main network structure in the future. Handover is needed when the mobile terminal moves from one cell to another in a DVB-H MFN or when the mobile terminal moves from one DVB-H SFN to another DVB-H SFN. Therefore, as DVB-H terminals go mobile, handover in the unidirectional DVB-H network becomes a critical issue.

There are differences between handover in DVB-H and handover in telecommunications networks such as UMTS. Take UMTS for example, the base station will communicate with the mobile station in the handover procedure and the availability of an interaction channel between the network and the terminal is essential for the successful completion of the handover. DVB-H networks have no information as to who is using their services at a given time and where the terminal is possibly going to perform handovers. Since the DVB-H transmitter cannot obtain information from the DVB-H terminals, the DVB-H terminals themselves based on their own decisions must perform the handover. Thus the main challenge for handover in DVB-H is that the handover in DVB-H can only be initialised and completed by the terminal itself without interaction with the transmitters.

Time slicing is the transmission mode used in DVB-H where the different services are transmitted using instantaneous high bit rates (up to 10 Mbps or more) [160] in different time slices. With the introduction of the time slicing mode in DVB-H, the DVB-H terminal can make handover measurements in the off burst time without service interruption [94]. The off burst time is the time interval in time slicing mode when the terminal is in sleep mode (refer to Fig. 1.4). Soft handover in DVB-H means that the service is not interrupted from the user's point of view. Thus handover in DVB-H is a soft handover. Time slicing in DVB-H is designed to save battery power and is not possible in DVB-T [95] because DVB-T uses continuous transmission.

1.3 Handover in Converged Networks

As converged network is becoming more and more important. The handover in the converged networks is also becoming critical. On one hand, the operators

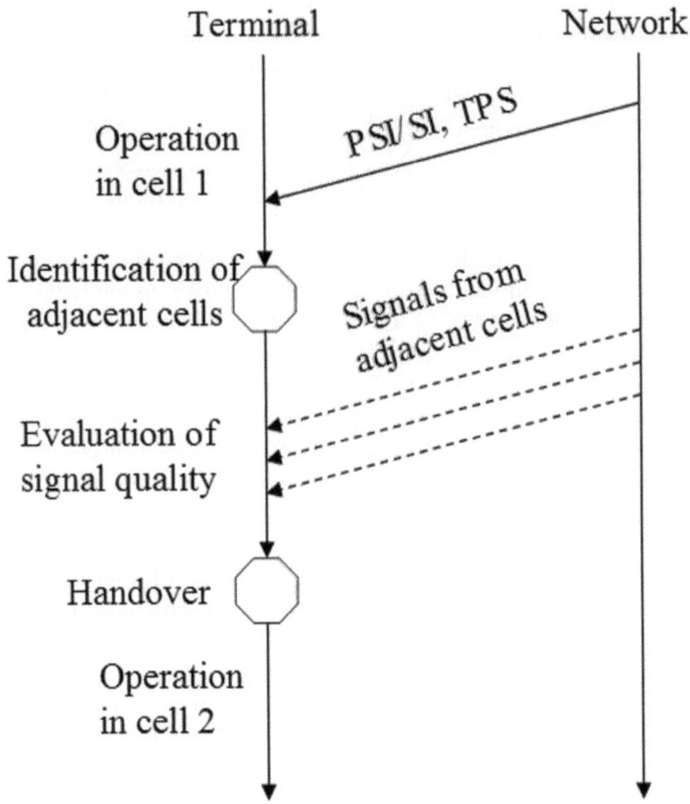

Fig. 1.1. Passive Handover According to [94]

want to integrate the new DVB-H service with the 3G streaming service that they have already set up, in order to save cost and to provide more interactivity to the users. On the other hand, the users do not care which technology is behind the service that they are consuming, they are caring more about the quality of the service that they received. All these issues pushed the converged network into the schedule.

A typical problem in such a converged network is the handover between DVB-H and UMTS. This kind of handover is a kind of radio resource allocation method. The so called intersystem soft handover can reallocate the radio resources between the DVB-H network and the UMTS network according to the trade-off between the operating costs and the quality of service of the network. UMTS is expensive in providing the same service to many different users because of its unicast mode. The UMTS multicast mode MBMS is only suitable for light-duty multimedia traffic because of the low bit rate compared to DVB-H (64kbps in MBMS is viable but 128 kbps or higher is problematic [132]. 256kbps for DVB-H was already used in the Berlin pilot trial operation [163] and 384kbps for DVB-H was already used in the Vienna pilot trial operation [63]). On the other hand, UMTS can provide interactivity to users which the DVB-H cannot provide. Thus UMTS offers better quality in case of interactivity. Therefore, there is a trade-off between the operating costs and the quality of service in the converged networks between UMTS and DVB-H.

Stochastic trees are extensions of decision trees that facilitate the modelling of temporal uncertainties [19, 21]. In its simplest and most useful form, a stochastic tree is a transition diagram for a continuous-time Markov chain, unfolded into a tree structure. The stochastic trees concept is utilized in the intersystem handover algorithm presented in Chapter 10 of this book.

1.4 Handover in Hybrid Broadcast Networks

DVB-H is not the only mobile broadcast standard in the world. There are different digital mobile broadcast standards being developed and emerging around the world, such as Terrestrial - Digital Multimedia Broadcasting (T-DMB) [4] from South Korea, Media Forward Link Only (MediaFlo) [149] from the USA and the Digital Multimedia Broadcasting - Terrestrial (DMB-T) from China. With the globalization, it becomes easier for people to move between different countries, which also makes the demand for watching the services in different countries increase. Thus it is necessary to have one device which can work with different broadcast standards. It is also a fact that different broadcast standards begin to coexist within one country. In this case, handover in the hybrid broadcast networks becomes also an issue. Such handover refers to the handover between different broadcast technologies and it happens when the user moves from one broadcast network (e.g. DVB-H) to another different broadcast network (e.g. DMB-T) or when the users lose the signals from one broadcast network and have to continue the service from another broadcast network. The handover between DVB-H and DMB-T in such a hybrid broadcast networks will be discussed in Chapter 11 of this book.

1.5 Passive Handover and Active Handover in DVB-H

Although some convergence terminals have both DVB-H and telecommunication capabilities, it is not always reasonable for a terminal to get in contact with the network to perform handover. Handover without an interaction channel in DVB-H is called passive handover while handover in DVB-H utilising an interaction channel such as a UMTS return channel is called active handover [66]. Illustrations of these two kinds of handovers are given in Fig. 1.1 and Fig. 1.2 [133]. Fig. 1.1 shows passive handover in DVB-H. This is the case for handover in dedicated DVB-H networks. In Fig. 1.1, the terminal receives the

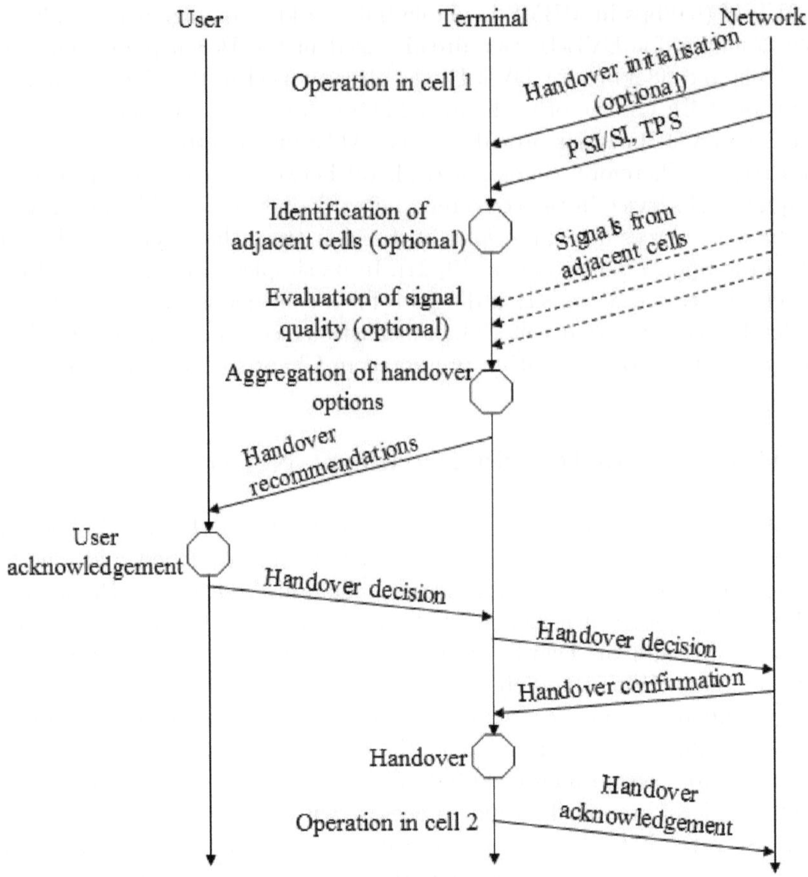

Fig. 1.2. Active Handover According to [94]

signalling information within the DVB-H transport streams, it then makes the signal measurement from adjacent cells when it reaches the cell border area. When the handover decision is made, the terminal will synchronize to the target cell signal and finishes the passive handover process. Fig. 1.2 shows the active handover process in DVB-H. This is the case when the DVB-H terminal uses an uplink channel (like UMTS) which it used in DVB-H handover. As shown in Fig. 1.2, the terminal will still receive the DVB-H signalling information from the transport streams. After the terminal made the handover decision it will inform the user of its handover recommendation, the user will either agree or reject the handover recommendation of the terminal. Once the users' acknowledgement is delivered by the terminal, the terminal will then inform the network through the uplink channel and perform or reject the handover after handover confirmation from the networks. Since DVB-H does not require a mandatory interaction channel, in this book passive handover in DVB-H terminals will be the main focus for handover in dedicated DVB-H networks where the terminal has no interaction channel with the network infrastructures.

1.6 Soft Handover in DVB-H

Soft handover is not possible for single antenna DVB-T terminals because there is no time interval gap for performing soft handover in DVB-T's continuous transmission mode. The DVB-H standard brings the possibility of soft handover for single antenna terminals. There are two main features that make soft handover possible in the DVB-H standard, one is time slicing and the other one is the mandatory cell id identifier [134]. Time slicing creates off times that can be used for the monitoring of the adjacent cells without interruption in the service consumption. Mandatory cell id identifiers assist the handover decision process and reduce the tuning failure probabilities.

Fig. 1.3 shows the basic soft handover scenario in DVB-H networks. Fig. 1.3 also shows the basic DVB-H network structure. In Fig. 1.3, the service application provider is in charge of the services application creation. The playout server is in charge of the service management. The DVB core network is the transmission network that delivers the services to the radio access networks. IP network is the best candidate for the transmission network. Region 1 and Region 2 shown in Fig. 1.3 are two different DVB-H Single Frequency Networks (SFNs) that are formed mainly by DVB transmitters and repeaters. The frequencies of the two single frequency networks are different so the DVB-H receiver will perform frequency handover when it moves from one of the single frequency networks to another. Fig. 1.3 also shows two different service streams. It can be easily seen that when the DVB-H receiver moves between Region 1 (SFN) and Region 2 (SFN), the DVB-H receiver receives the same service stream by the aid of handover.

More detailed technical information about DVB-H is provided below.

1.7 Technical Features of DVB-H

Time slicing, Multiprotocol Encapsulation-Forward Error Correction (MPE-FEC), the 4K mode, the in-depth interleavers, DVB-H signalling (including the mandatory cell id identifier) and the use of 5 MHz bandwidth are the essential elements that are introduced in DVB-H. These features are located in the data link layer and the physical layer of the DVB-H protocol stack. Time slicing (in the data link layer) and DVB-H signalling (in the data link layer and the physical layer) are the two features that are directly related to DVB-H handover.

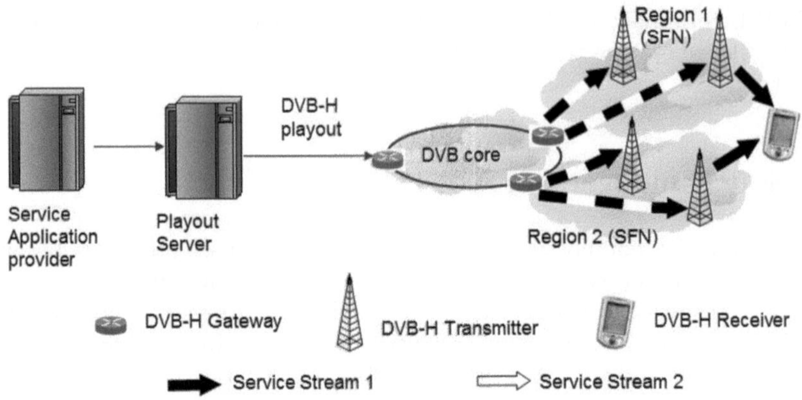

Fig. 1.3. Soft Handover Illustration

1.7.1 DVB-H Protocol Stack

The DVB-H protocol stack is shown in Fig. 1.4. The newly introduced DVB-H technical features are in the data link layer and the physical layer. The application services may be sent via RTP (Real Time Protocol) [129] for real time content (for example a TV program). Non-real-time data maybe sent via FLUTE/ALC (File Delivery Over Unidirectional Transport/Asynchronous Layered Coding) [130] data carousel (for example for file downloads). The Electronic Service Guide (ESG) is also broadcasted using FLUTE/ALC. The handover issue in DVB-H is mainly an issue for the data link layer and the physical layer. An analysis and simulation of DVB-H link layer is done using finite-state Markov models in [164].

Application Layer	Real Time Content	File Based Content	ESG	
Presentation Layer	Source Coding H.264(Mpeg4)	Source Coding H.264(Mpeg4)	Coding Encapsulation (XML)	
Session Layer	RTP	FLUTE/ALC		
Transport Layer	UDP			
Network Layer	IP (IPv4/IPv6)			
Data Link Layer	MPE (Time Slicing)			SI/PSI
	MPEG-2 Transport Stream			
Physical Layer	TPS	DVB-T		

Fig. 1.4. DVB-H Protocol Stack

1.7.2 Time Slicing

Time slicing is illustrated in Fig. 1.5. Time slicing is used in DVB-H to transmit data in periodic bursts with significantly higher instantaneous bit rates compared with the bit rates used if the data are continuously transmitted as in DVB-T.

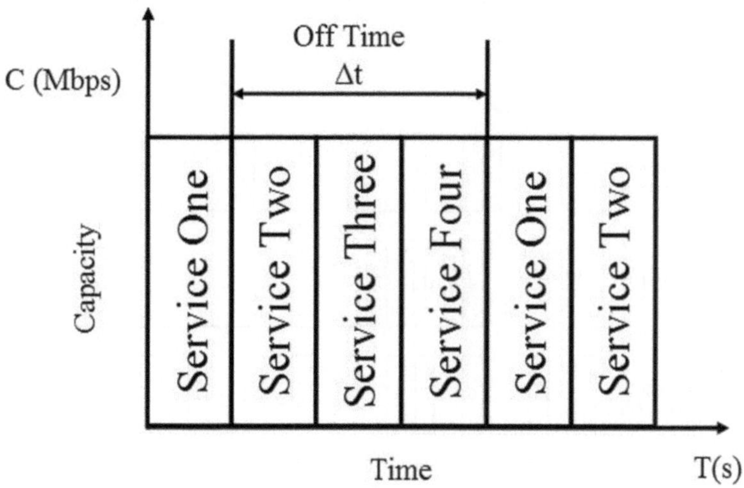

Fig. 1.5. Time Slicing Illustration

Time slicing is in some aspects similar to the TDMA technology used in GSM standards [86]. The difference between the TDMA in GSM and the time slicing in DVB-H are: TDMA in GSM has fixed duration for each time slot while the time slot duration in time slicing could be variable; TDMA in GSM assigns time slots to different users while the time slots in time slicing are assigned to different transmitted services; the TDMA in GSM assigns time slots to both downlink and uplink channels while the time slots in time slicing are assigned to downlink channels only. Time slicing enables the tuner in a DVB-H receiver to stay active only a fraction of the time, while receiving bursts of a requested service, this saves battery power. It is claimed that up to 95% power saving can be achieved compared with conventional and continuously operating DVB-T tuners [134]. The high bit rate signals will be buffered in the receiver memory. A brief performance analysis of the time slicing scheme in DVB-H is done by simulation in [101]. A multi-antenna diversity approach is extremely difficult because of space limitations [133]. Time Slicing offers, as an extra benefit, the possibility to use the same front-end to monitor neighboring cells between bursts, making seamless soft handover possible. [102] showed how the off times between the transmissions bursts can be used to perform handovers, how they have to be synchronized and what boundary conditions exist. A technology called "phase shifting" is proposed as a solution. In the "phase shifting" proposal, a static phase shift exists between the two neighboring cells shown in Fig. 1.6. Fig. 1.6 shows how the overlapping of IP packets (one example marked in grey in the figure) ensures loss-free handovers [68]. The main idea of "phase shifting" is that the phase shift between neighboring two cells should be at least the maximum time of the time slice plus the time the terminal needs for synchronization to the new signal in the handover target cell.

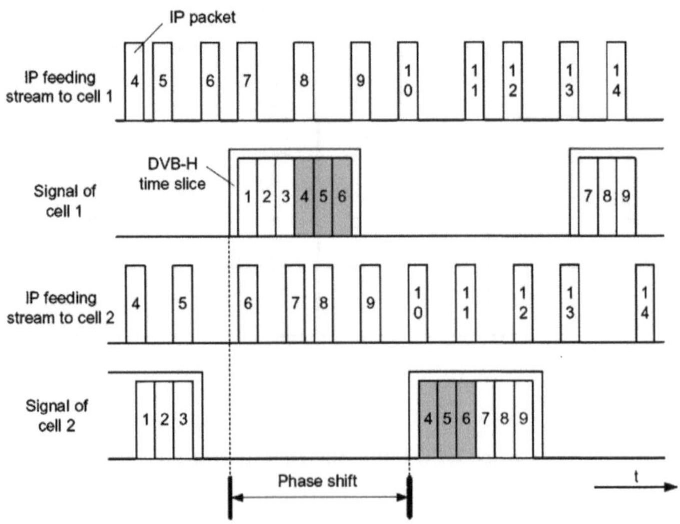

Fig. 1.6. Phase Shifting Principle from [68]

Fig. 1.7 shows the comparison of DVB-T transmission and DVB-T/H combined transmission. It is possible to use a combination of DVB-H (time-sliced) and DVB-T (not time sliced) services in a single multiplex as shown in Fig. 1.7 [94]. This kind of combination is necessary to incorporate the DVB-H services into the existing DVB-T infrastructure. However, the power saving is decreased in this case due to a smaller data rate being available for time sliced services [133]. In the DVB-T transmission shown in Fig. 1.7, the five services are transmitted together in the transport stream. In this case, the terminal has to receive all five services before decoding the service it targets to. In the combined transmission of DVB-T and DVB-H shown in Fig. 1.7, although the DVB-T terminal has to do the same as in the pure DVB-T transmission, the DVB-H terminal needs only to decode the equivalent of three services when it uses time slicing mode. In this way, the DVB-H terminal saves battery power by using time slicing mode.

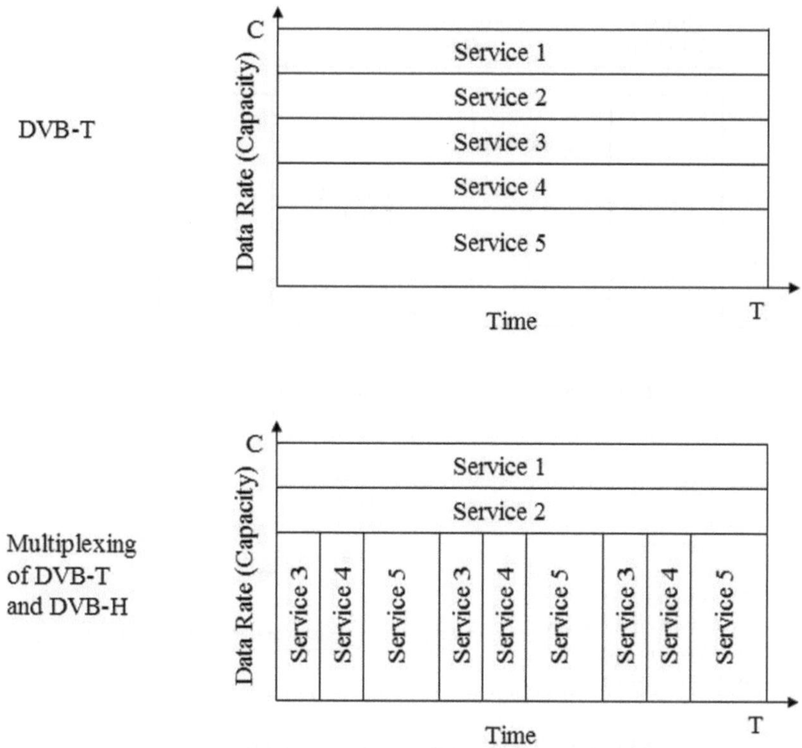

Fig. 1.7. DVB-T and Multiplexing of DVB-T and DVB-H

Another benefit of time slicing in DVB-H is that it is unique in terms of the power saving achieved. This means that the amount of power savings achieved by time slicing in DVB-H could not be obtained when time slicing is used in DAB or DMB [144].

Depending on the transmission bit rate, burst size and burst duration, the off time Δt in the transmission stream can vary [104]. According to [68], the burst parameters are shown in Fig. 1.8 and the formulas used to calculate the length of a burst, the off time and the achieved saving in power consumption are shown in Fig. 1.9. The DVB-H receiver can use this off time to synchronize and initialize soft handover to another cell that would be impossible without the use of time slicing.

1.7.3 MPE-FEC

Multi-Protocol Encapsulation (MPE) is a method to transmit IP data over DVB networks [74]. It specifies the carriage of IP packets within MPEG Private Data sections. DVB-H is designed to use the broadcasting frequency

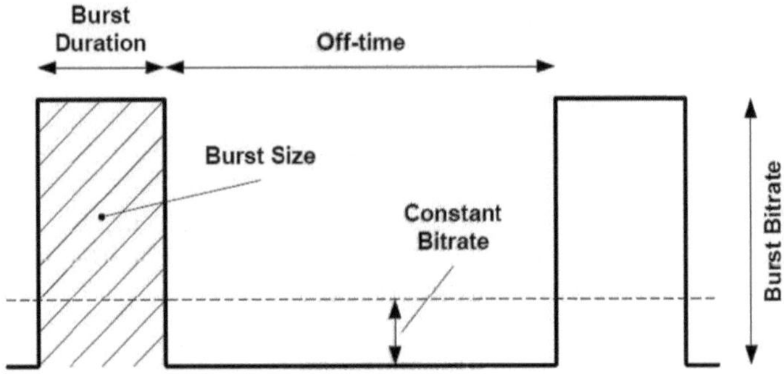

Fig. 1.8. Time Slicing Burst Parameters [68]

Bd	Burst Duration (seconds)
Bs	Burst Size (bits)
Bb	Burst Bitrate (bits per second)
Cb	Constant Bitrate (bits per second)
Ot	Off-time (seconds)
St	Synchronization Time (seconds)
Ps	Power Saving (per cent)
Dj	Delta-t Jitter (seconds)

$$Bd = \frac{Bs}{Bb \cdot 0.96}$$

$$Ot = \frac{Bs}{Cb \cdot 0.96} - Bd$$

$$Ps = \left(1 - \frac{(Bd + St + (3/4 \cdot Dj)) \cdot Cb \cdot 0.96}{Bs}\right) \cdot 100\%$$

Fig. 1.9. Time Slicing Burst Parameters [68]

bands in Very High Frequency (VHF) and Ultra High Frequency (UHF) frequency spectrum. The mobile reception of DVB-H is characterized by Non-Line of Sight, multipath, Doppler impairments, strong propagation loss (especially for indoor reception), poor receiving antenna gain [100], and mobile channel interferences from adjacent TV and GSM channels and environmental factors like weather and traffic. As a result, accessing a downstream high bitrate service with a handheld terminal is very demanding. The objective of the Multi-Protocol Encapsulation Forward Error Correction (MPE-FEC) is to improve the Carrier/Noise (C/N) ratio and Doppler tolerance in mobile channels and to improve the tolerance to impulse interference [94]. However, MPE-FEC only works within individual time slices [145] because the size of one time slicing burst exactly corresponds to the content of one MPE-FEC frame [134]. Consequently, if a single transmission error cannot be corrected, the service drops out not only for the duration of the burst but also during the time up until the next burst is received. Fig. 1.10 shows how MPE-FEC works. As illustrated in Fig. 1.10, using MPE-FEC the IP datagrams are stored in the "Application Data Table" and the Reed Solomon (RS)encoder RS (255, 191) is applied to each Application Data Table row to produce 64 byte FEC code words. The contents of both the Application Data Table and the RS data table are transmitted column by column. As the RS data table is read column by column, each of the RS data table bytes adjacent in a row are now separated by a distance equal to the number of table rows. This provides the receivers with the ammunition to fight against mobile channel impairments [160].

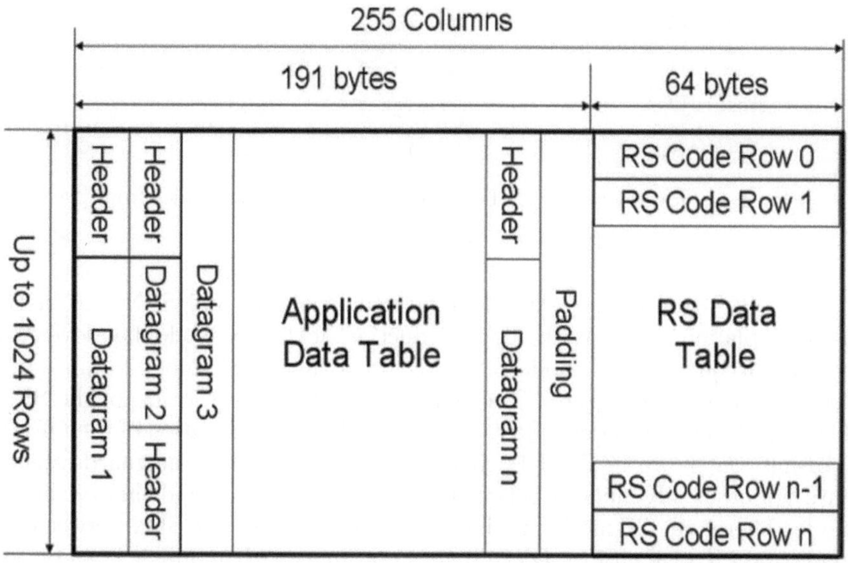

Fig. 1.10. MPE-FEC Illustration According to [160]

Handover usually happens in the cell border area where the signals from different transmitters are very weak thus the reception is vulnerable to impulse interferences. In addition the handover usually happens in mobile terminals. So MPE-FEC helps handover by improving the C/N ratio and Doppler tolerance in mobile channels and by improving the tolerance to impulse interferences.

1.7.4 4K Mode and In-depth Interleavers

The 4K mode and the in-depth interleavers affect the physical layer of DVB-H but do not affect the soft handover directly. However, their objectives are to improve Single Frequency Network (SFN) planning flexibility and to protect against short noise impulses caused by, e.g. ignition interference and interference from various electrical appliances [94, 146]. In this case, they affect the mobile reception of DVB-H signals. 4K together with 2K and 8K are referring to the number of subcarriers used in DVB-H OFDM transmission mode. The parameters for the different DVB-H OFDM transmission modes are shown in Fig. 1.11.

OFDM parameter	Mode 2K	4K	8K
Overall carriers (= FFT size)	2048	4096	8192
Modulated carriers	1705	3409	6817
Useful carriers	1512	3024	6048
OFDM symbol duration (μs)	224	448	896
Guard interval duration (μs)	7,14,28,56	14,28,56,112	28,56,112,224
Carrier spacing (kHz)	4.464	2.232	1.116
Maximum distance of transmitters (km)	17	33	67

Fig. 1.11. Parameters of DVB-H OFDM Transmission Modes from [147]

The 4K mode offers a trade-off between mobility and SFN size in the network planning [94]. With the introduction of the 4K mode, handover will be more frequent in 4K mode compared with that in 8K mode because of a decrease in the cell size [70]. The cell size here refers to the size of the SFN. The size of the SFN is dependent on the guard interval duration of the OFDM transmission shown in Fig. 1.11. It can be seen easily that the guard interval duration of the 4K mode is between that of the 2K mode and that of the 8K mode. Therefore, 4K mode provides a kind of medium size SFN between 2K mode and 8K mode. Since DVB-T does not include 4K mode, it is an option only in dedicated DVB-H networks [68]. While for the 2K and 4K modes the in-depth interleaver increases the flexibility of the symbol interleaving by decoupling the choice of the inner interleaver from the transmission mode used [94, 68]. This means that the 8K mode interleaver buffer can be used in the 2K and 4K modes as an alternative interleaver. Furthermore, the use of the 8K interleaver provides the flexibility to switch between 2K, 4K and 8K modes without changing the interleaver. The formation of the in-depth interleaver is shown in Fig. 1.12. The in-depth interleaver makes use of the available memory for the 8K mode and improves the performance of the 2K and 4K modes. This kind of using alternative large memory available for 8K mode for the use of 2K and 4K mode is called in-depth interleaving [147]. Using an in-depth interleaver in the 2K and 4K modes can further improve the terminal's robustness in mobile environments and impulse noise conditions because of the alternative large memory being used.

1.7.5 DVB-H Signalling

The objective of DVB-H signalling is to provide robust and easy-to-access signalling to DVB-H receivers, thus enhancing and speeding up service discovery [94]. It should be noted that DVB-H is based on DVB-T and most of the DVB-H specifications in the physical layers are the same as those of DVB-T

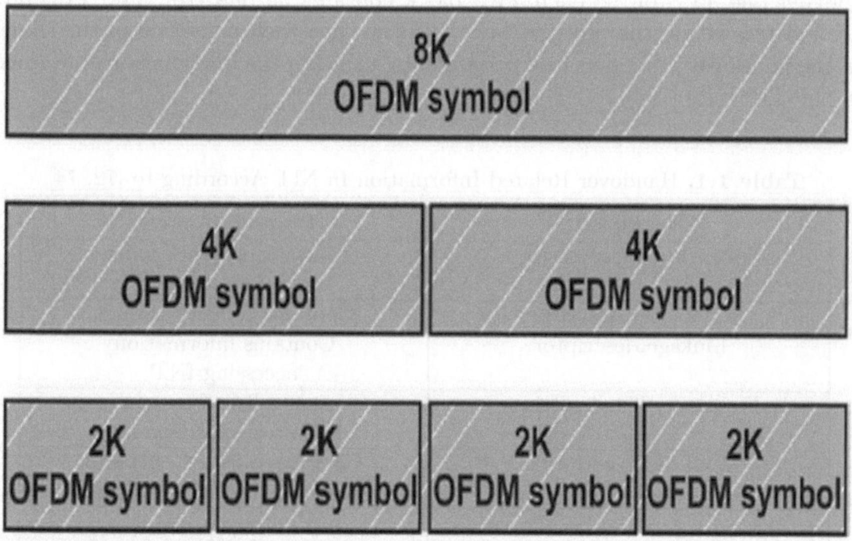

Fig. 1.12. In-depth Symbol Interleaving of OFDM Symbols from [147]

that can be found in [70]. Besides the specifications in common with DVB-T, DVB-H has unique physical specifications. Only the DVB-H signalling used for handover is considered in this section. The signalling bits specified for DVB-H but not used directly for handover will not be illustrated. There are two kinds of signalling information the DVB-H receiver can use. One is Transmission Parameter Signalling (TPS) signalling bits in the physical layer. The other is DVB-H specific signalling within Program Specific Information (PSI)/Service Information (SI) [72, 74, 69] that forms a part of the DVB-H transport streams. In [71] Service Information (SI) is referred to as Program Specific Information (PSI). There are also "reserved for future use" bits available for future parameter additions. The unused bits are ignored by the receivers. PSI/SI is the core signalling for enabling service discovery within the DVB systems. Since the PSI/SI used within DVH-H is different to that of other DVB systems, a subset of PSI/SI for IPDC over DVB-H is defined in [87]. The PSI/SI data enables a DVB-H receiver to discover IPDC over DVB-H specific services in the transport stream [72] and also provides essential information for enabling handover.

(In order to implement handover in DVB-H, the receiver needs to receive signalling information from the network. For illustration purposes the handover related parameters contained in the Network Information Table (NIT) of the PSI/SI table are derived from [72, 74] and are shown in Table 1.1.)

The Transmission Parameter Signalling (TPS) is defined over 68 consecutive OFDM symbols, referred to as one OFDM frame. Each OFDM symbol conveys one TPS bit so each TPS block contains 68 bits [70]. The TPS bits are located within the physical layer [146] so the synchronization of the signal in the handover process is first obtained by utilising the information contained in the TPS bits [105].

Table 1.1. Handover Related Information In NIT According to [72, 74]

Descriptor	Purpose/Content
Network_name_descriptor	Contains network name information
Service_list_descriptor	Contains services listings
Linkage_descriptor	Contains information accessing INT
Frequency_list_descriptor	Contain a list of frequencies for a transport stream
Cell_list_descriptor	Contains a list of cells and subcells including their coverage areas
Cell_frequency_link_descriptor	Contains a list of cells and frequencies used for the transport streams
Terrestrial_delivery_system_descriptor	Contains information about the centre frequency, bandwidth, code rate, etc.
Time_slice_FEC_identifier_descriptor	Contains information about the time slicing and MPE-FEC being used

The Synchronization Word bits aid the receiver in synchronizing with the target transport stream and/or frequency. The Cell Identifier conveys unique cell identification information to the receiver. The PSI/SI table provides information on the DVB-H services carried by the different transport streams. Handover related information in the PSI/SI table is mainly contained in the NIT (Network Information Table), the PAT (Program Association Table), PMT (Program Map Table) [72] and INT (IP/MAC Notification Table) [74]. The details of the PSI/SI tables will be presented in Chapter 4.

Using the TPS bits and PSI/SI tables contained in the received transport streams the DVB-H receiver can initialize and perform the handover efficiently.

1.7.6 5 MHz Bandwidth

DVB-T standards use the 6 MHz, 7 MHz or 8 MHz raster in the frequency spectrum (namely UHF and VHF). The introduction of 5 MHz bandwidth into DVB-H provides new possibilities for using frequency spectrum other than that allocated to traditional broadcast use, for example in L band, which creates new challenges in terms of receiver design and also provides benefits in terms of system performance such as tolerance to Doppler shift in a mobile environment [68].

1.8 DVB-H System Components

The basic functional Components for a DVB-H system are shown in Fig. 1.13. One functional component can be several equipments. On the other hand, several functional components can be combined into one equipment. In Fig. 1.13, the concrete arrows stand for the information link which are compulsory for a DVB-H system, while the dotted arrows stand for the optional link for a DVB-H system. The details of the involved components are described below:

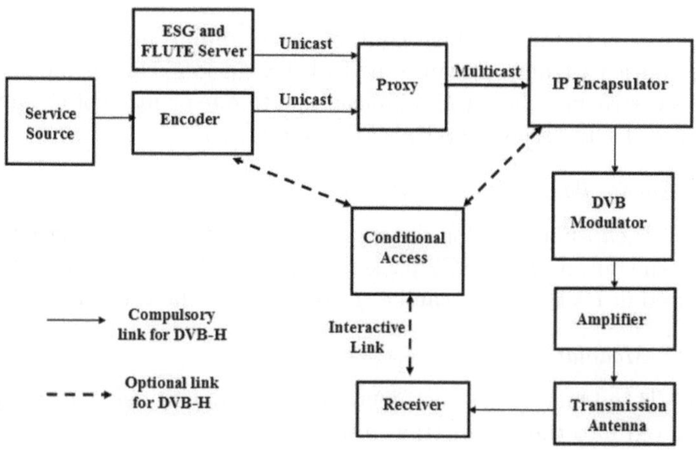

Fig. 1.13. Basic Functional Components for DVB-H System

A. Service Source

The service source is the component located on the side of the application providers. The service source produces different audio/video/data programs either encoded in a certain format or not encoded. Some of the programs could be the programs that are already available as the home TV received channels and the other of the programs could be the new channels which are targeting specifically the mobile portable devices. It is a very important component for the service providers or the content providers to re-sell their existing programs to the consumers.

B. Encoder

The encoder is the component which encodes the contents from the service source node to the suitable audio/video/data format that can be used by the DVB-H systems. The usually used video format for DVB-H is H.264, the usually used audio format for DVB-H is HE AAC and the data format is usually XML. Other audio/video/data formats are also optional.

C. ESG and FLUTE Server

The ESG and FLUTE Server are used to produce the ESG and convert it to the correct format for transmission in the FLUTE Carousel. The FLUTE Server can also be used to transmit other normal data files (e.g. softwares for update).

D. Proxy

The proxy is the node which convert the incoming unicast traffic to the outgoing multicast traffic. Because the FLUTE session is made of one or more IP multicast streams, the terminal must tune to one or more of the multicast streams in order to receive the FLUTE session [91].

E. IP Encapsulator

The IP Encapsulator is for the encapsulation of the incoming IPv4/IPv6 packets to the outgoing MPEG2 packets. These MPEG2 packets are the same as those used in DVB-T, thus can be transmitted by the DVB-T transmitters.

F. DVB Modulator

The DVB modulator has the function of modulating the MPEG2 packets to be radiated into the air for transmission.

G. Amplifier

The amplifier is in charge of the amplification of the modulated DVB signals. It has the amplification power ranging from several Watts to several Kilo Watts.

H. Transmission Antenna

The transmission antenna has also various size depending on the intended transmission power and the targeted coverage area. It can be either indoor or outdoor antennas.

I. Receiver

The DVB-H receiver is a device that is capable of receiving DVB-H signals. Mobile phone with incorporated DVB-H receiving capability is a typical DVB-H device. PDAs, other mobile portable receivers and even laptops are also typical DVB-H devices.

J. Conditional Access

The Conditional Access component is optional in the DVB-H system. It is mainly for encrypting and billing the transmitted contents in the DVB-H network, thus it works as a content/service protection system. There are three content encryption methods available for the DVB-H system: ISMACryp (International Securities Market Association Encryption), SRTP (Secure Real-Time Transport Protocol) and IPsec (IP Security) [88]. The ISMACryp and the SRTP normally works on the Encoder side, while the IPsec normally works on the IP Encapsulator side. Furthermore, different level of keys are delivered to the receiver using either broadcast link or telecommunications link (if available) for decrypting and charging of the service. When a DVB-H system has no Conditional Access or the Conditional Access does exists but is not being used, i.e. the service is unscrambled, the transmitted service is called Clear-To-Air or Free-To-Air. When the DVB-H system has a functional Conditional Access (i.e. the service is scrambled) but no billing function implemented, it is called Free-to-View.

1.9 Book Structure

Chapter 2 presents the motivation of the DVB-H research including why should we look at the problem of handover in DVB-H; what are the main challenges in the handover process in DVB-H; how should we do the research for the handover in DVB-H including the different approaches being used to address the challenges in DVB-H handover; how should we evaluate the handover algorithms in handover for DVB-H.

Chapter 3 provides a comprehensive survey of the research that has been conducted on the handover issues in DVB-H networks. Different research results are briefly presented. Different research projects related to the DVB-H handover are also presented.

Chapter 4 presents the DVB-H signalling information, including the PSI/SI tables and the TPS bits. The methods about how to obtain and analyze the DVB-H signaling information are also presented.

Chapter 5 introduces the Electronic Service Guide (ESG) concepts used in DVB-H. As there are two categories of ESG at the time: DVB IPDC ESG and OMA BCAST ESG, these two different ESGs are presented in detail.

Chapter 6 focuses on the different handover algorithms for dedicated DVB-H networks in general. In the dedicated DVB-H networks, only the DVB-H receiver is considered, no matter whether the receiver has addition 3G or other uplink capability incorporated.

Chapter 7 presents a handover algorithm based on post processing of SNR values for a dedicated DVB-H network. The performance analysis of the algorithm is analyzed using both theoretical analysis and simulation via MATLAB and OPNET. The focus in this chapter is how to choose the right handover measurement criteria.

Chapter 8 presents a repeater aided handover algorithm for a dedicated DVB-H network. The repeaters are widely used in the radio communication networks. This chapter shows how the repeaters can be used for the handover in DVB-H.

Chapter 9 provides the soft handover probability calculation of the repeater aided handover algorithm. The calculation are expressed using both mathematical equations and computer simulation. This chapter shows how the calculation results are related to the battery power saving benefits for DVB-H receivers.

Chapter 10 discusses the handover in the converged broadcast and telecommunication networks by using the converged DVB-H and UMTS network as an example. The stochastic trees model is introduced and being used for the intersystem handover between DVB-H and UMTS within the converged networks. Analysis and simulation results are given to show how the stochastic trees model can be used in the intersystem handover process in the converged networks.

Chapter 11 discusses the handover in the hybrid broadcast networks. The focus of the chapter is on the handover algorithms between DVB-H and one of the Chinese digital broadcast standards: DMB-T. It also shows the importance of such handovers in the real application scenarios.

The future research directions for DVB-H and DVB-H handover are discussed and conclusions are presented in Chapter 12.

Problems

1.1. Why was DVB-H developed?

1.2. How does DVB-H make impact on the telecommunication and the broadcast world?

1.3. What are the new features of DVB-H compared with DVB-T?

1.4. Why should we consider handover problem in DVB-H?

1.5. What are soft handover, passive handover and active handover in DVB-H?

1.6. What are the basic components for a DVB-H system?

1.7. What are Clear-To-Air (Free-To-Air) and Free-To-View?

2

Motivation and Approaches

2.1 Motivation

This section presents the context that created handover issues in DVB-H networks. Before talking about the origin of the handover issues in DVB-H it is necessary to take a look at the services that are transmitted in DVB-H networks. The service contents in DVB-H networks are delivered in the form of IP-packets using IP-based mechanisms or in the form of other network layer datagrams encapsulated into Multi Protocol Encapsulation-sections (MPE-sections) [67, 134]. This kind of service is called IP Datacast (IPDC) [53]. IPDC was developed by the DVB ad-hoc group Convergence of Broadcast and Mobile Services (CBMS) [53] over the DVB-H standard [93]. The IPDC over DVB-H standard complements the DVB-H standard by defining OSI layers 3-7 and refining some of the OSI layer 2 specific protocols, especially Program Specific Information (PSI) and Service Information (SI). Although IPDC services can be offered via existing GPRS or UMTS cellular networks by MBMS [135, 96], MBMS is only suitable for light traffic services such as short video clips. For a heavy duty service like streaming using H.264 video encoding, QVGA format, 25fps and 384kbps, DVB-H is a better solution. This is largely because of that DVB-H is a broadcasting technology with up to 10 Mbps bit rates depending on the modulation parameters [160]. IPDC brings new characteristics for DVB-H networks. The benefits are as follows [68]:

1. IPDC provides a platform for true convergence of services between DVB-H and cellular telecommunication networks (GPRS/UMTS).
2. IPDC allows the coding to be decoupled from the transport layer, that is, all the different coding techniques can be used above the UDP/IP (User Datagram Protocol/Internet Protocol) layer, thus opening the door to a number of features benefiting handheld mobile terminals including a variety of encoding methods, which only require low power from a decoder (Decoding high bandwidth MPEG-2 encoded streaming video/audio is relatively power consuming).

3. IPDC is relatively insensitive to any buffering or delays within the transmission (unlike MPEG-2), this is because IPDC is utilizing IP protocol which has long been developed upon the non-quality-assured Internet environment.
4. IPDC is well suited for time-sliced transmission.

Because the mobile handover environment in DVB-H is as challenging as the non-quality-assured Internet environment, an IPDC service is suitable for handover in DVB-H.

VALIDATE and MOTIVATE are two European projects that addressed the issues of mobile reception of DVB-T signals [2]. Laboratory tests and field trials have shown that mobile applications of DVB-T are feasible using the code rate = 1/2 modes of the specification [136] which uses more resources than the code rate 3/4. But tests of receivers also showed the limits of performance achievable for mobile television without enhancements to the receivers [97]. In addition, the power consumption of mobile reception of DVB-T is a big issue for battery powered terminals [3, 98].

DVB-H is the technique that was rolled out for the mobile portable reception of IPDC contents [137, 142]. Similar standards are used in Japan and Korea for mobile data broadcasting [143, 4]. The achieved transmission bitrate is lower for mobile reception compared with that for static reception because of the challenging mobile environment. Therefore, the transmission power has to be higher to achieve an acceptable quality for user to view if an assumption is made that the high transmitted power will make it easier for the terminals to decode the received signals. It has also been shown that in general the maximum speed for mobile DVB-T reception in single frequency networks (SFN) is lower than in Multi-Frequency Networks (MFN) for the same set of parameters [99]. Low power DVB-H transmitters offer the possibility of multi-frequency cellular structured DVB-H networks for the broadcast of localized services. One network scenario is to co-locating the DVB-H and 3G UMTS transmitters where the cell size of DVB-H is usually smaller than that of the stand-alone DVB-H transmitters. Smaller cell size for DVB-H provides desirable opportunities for the provision of localized services. With smaller cell size, handover in DVB-H becomes a critical issue. Some research has been done about handover in multi-frequency cellular structured DVB-H networks [68, 104].

Handover in traditional cellular telecommunications networks (like GSM) refers to the mechanism that transfers an ongoing call from one cell to another as a user moves through the coverage area of the cellular system [38] and has long been a research topic. However, handover in DVB-H refers to the switching of the reception of IP based services from one transport stream to another when the terminal moves through the coverage area of a DVB-H network [67].

Soft handover is usually used to mean that radio links are added and removed in such a way that the device always keeps at least one radio link to the base station [106], thus no service interruption happens. In DVB-H, this means that the received frequency and/or transport stream is changed without interruption of the on-going reception.

A stationary DVB-H terminal can assume that a transport stream on a given frequency will be constantly available during its operation. However, DVB-H is mainly for mobile portable terminals and a mobile terminal will face the situation that the selected transport stream signal is no longer available on the tuned frequency if the terminal is moving out of the reception area. In order to continue the selected service, the mobile terminal then needs to automatically select and tune a different frequency carrying either exactly the same transport stream or a different transport stream containing the same service.

If the mobile terminal moved from one cell to another cell of the same network, the same set of transport streams could be available and if available they will be on different frequencies. The mobile terminal has to determine on which frequency the lost transport stream is transmitted in the entered cell [69].

If the mobile terminal moves from a cell belonging to one network into a cell of another network then the lost transport stream is not necessarily available. In this case the mobile terminal might want to find out if the service that had been selected before is still available on some transport stream of the entered network or if there are alternative services to select. Therefore, two situations exist:

1. If the previously selected service is still available, the mobile terminal needs to determine the transport stream that carries the service and the frequency of that transport stream in the entered network.
2. If the service is not available, the mobile terminal might try to select an alternative (which could be a local variation of the original service or an associated service) before it prompts the user for a decision. Deploying this mechanism, co-operating networks might provide automatic handover between services of similar program type or services that provide additional information such as traffic announcements.

Fig. 2.1 shows the two handover situations presented above: handover from one cell to another cell of the same network (within DVB core network 2) and the handover between cells of different networks (from DVB core network 1 to DVB core network 2). The DVB core network is usually an IP network. Such an IP network is owned or rented by a network operator. It is set up to transmit the services generated by the service application providers to the radio access networks (the IP-to-DVB encapsulators and the DVB-H transmitters). The DVB core 1 and DVB core 2 in Fig. 2.1 are not connected each other because each DVB core network is operated by a different network operator in the case shown in Fig. 2.1. Regarding these two different kinds of situations, from the application point of view, handover for DVB-H can be divided into Physical

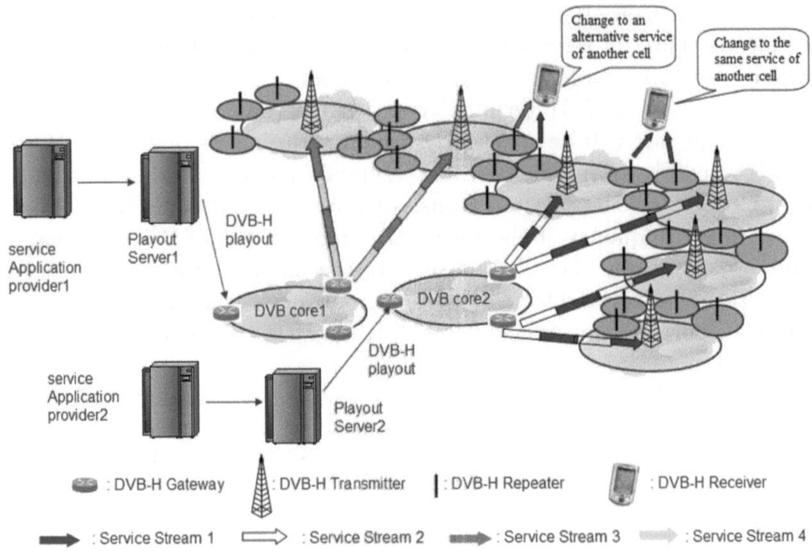

Fig. 2.1. General Handover Situations in DVB-H Networks

Handover and Service Handover depending on whether or not the service is changed.

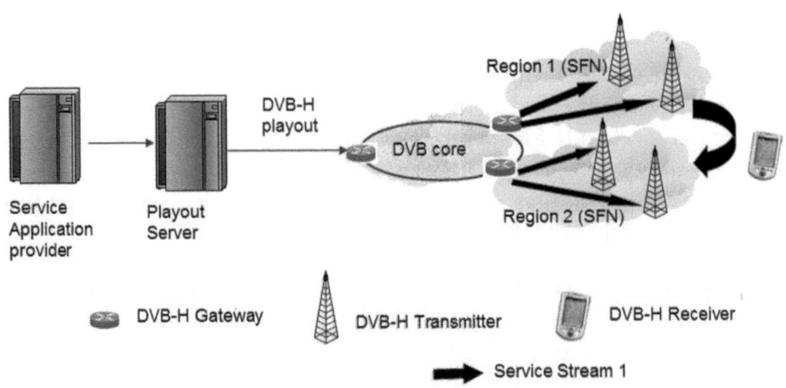

Fig. 2.2. Physical Handover Illustration

Physical handover is shown in Fig. 2.2. In this case, the terminal moves from one DVB-H cell to another without the received service being changed. However, the transmitter frequency sending the transport stream changes.

Service handover is shown in Fig. 2.3. In this case when the receiver moves from one cell to another, it begins to receive a transport stream that is different from the one it received in the original cell.

DVB-H is intended to carry IP data services. In order to provide diverse IP data services it is expected that a DVB-H cell will usually be smaller than a DVB-T cell [100]. There are also other reasons why a DVB-H cell is usually smaller than a DVB-T cell. They are low-gain DVB-H receiver antenna, low height of DVB-H transmitter antenna, building penetration losses for DVB-H indoor reception and fast fading in a mobile environment [100]. Thus, low power transmitters serving a network operating in a multi-frequency network (MFN) mode, in networks composed of one or more Single Frequency Network (SFN) areas or a mixture of these two topologies are the main network structure types for DVB-H.

One of the main factors that effect the selection of the network topology is the need for localized services. Depending on the density of the localized services, the network topology can vary from a MFN composed of single transmitter cells to networks composed of one or more SFNs. If localized services are not supported at all within the network, the whole network can consist of a single SFN where handover is not needed. All other network types, except the network type of a single SFN, result in handovers. The handover frequency in this book refers to the number of handovers required for a DVB-H receiver in its power-on duration. It is dependent on the size of the cells, terminal mobility, and environmental factors (such as rural or urban) which are similar to those in UMTS/GSM, etc. As low power transmitters serving a multi-frequency cellular structured network are expected to be a typical network structure for DVB-H, handover becomes a critical issue.

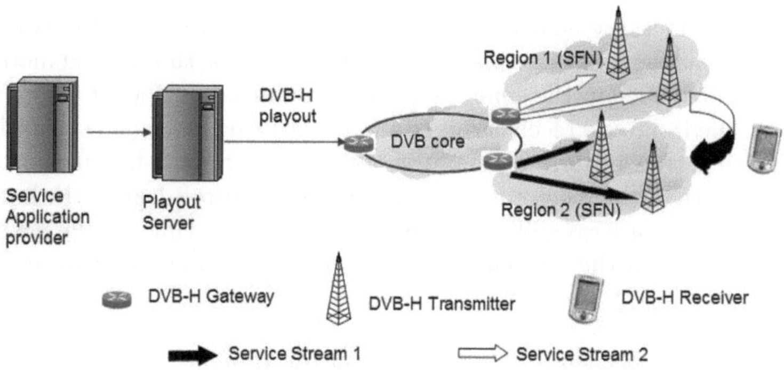

Fig. 2.3. Service Handover Illustration

Handover is the switching of a mobile signal from one channel or cell to another. For DVB-H, this chapter defines handover as a change of transport stream and/or frequency. In this chapter the two different terms "channel and transport stream" are used for the same meaning. They all mean the path along which a communications signal is transmitted.

The handover schemes for DVB-H are soft handover because of the existence of time slicing.

It must be noted that the handover categories may overlap. For example a soft handover can be both a service handover and an active handover at the same time.

2.2 Approaches

There are different ways to address the handover problem in DVB-H. Some of the handover algorithms in other networks, such as the handover in telecommunication networks and the handover in mobile IP network, can even be borrowed and taken as a reference in the DVB-H networks. The handover issue for DVB-H in this book is approached from two aspects: the handover stages of a DVB-H handover and the handover challenges that need to be addressed for a DVB-H handover.

2.2.1 Handover Stages

Handover in DVB-H consists of three stages: handover measurement, handover decision-making based on the handover criteria, and handover execution [112]. All the previous research work on handover in DVB-H can be categorized into or was targeting these three stages.

A. Handover Measurement

Handover measurement is the first of the handover stages. In DVB-H the handover measurement takes place in the off time of the time slicing mode. The terminal will switch off the tuner and the demodulator in the off burst period. However, the front-end receiver has to keep measuring the signal strength from neighbouring transmitters to monitor the signal strength fluctuation. If the signal strength of the received signal is degraded to some degree, the handover decision-making process will be triggered. The wake up time for the next burst will be signalled in the current burst period. The detailed procedure is given in [104].

B. Handover Decision-making Based on the Handover Criteria

In the second stage of the handover process, the DVB-H terminal will decide whether it should perform handover based on the pre-defined handover criteria. The most commonly used handover criteria are the Received Signal Strength Indicator (RSSI)and the Signal Noise Ratio (SNR) [104, 106]. When the RSSI or the SNR is identified as degraded to some degree from the handover measurement of the first stage, the handover decision-making process will be triggered. Taking the SNR as the handover criteria for example, once the SNR threshold margin value *s_th* is reached for a certain threshold time *t_th*, the receiver will tune to the frequency with the strongest SNR value to continue service reception. The SNR and the duration threshold is shown in Fig. 2.4. Fig. 2.4 was obtained from the simulation conducted in [106]. In Fig. 2.4 the DVB-H terminal is receiving the signal from subcell 1 from the beginning. At time 12 the terminal begins measuring the signal from subcell 2. After time duration t_th the terminal will handover to the signal from subcell 2 at around time 13.5. In addition to the physical layer parameters, it is increasingly important to take the quality of currently received IP streams as one handover criterion especially within a MFN network. This is also recognized in [42] and in [68].

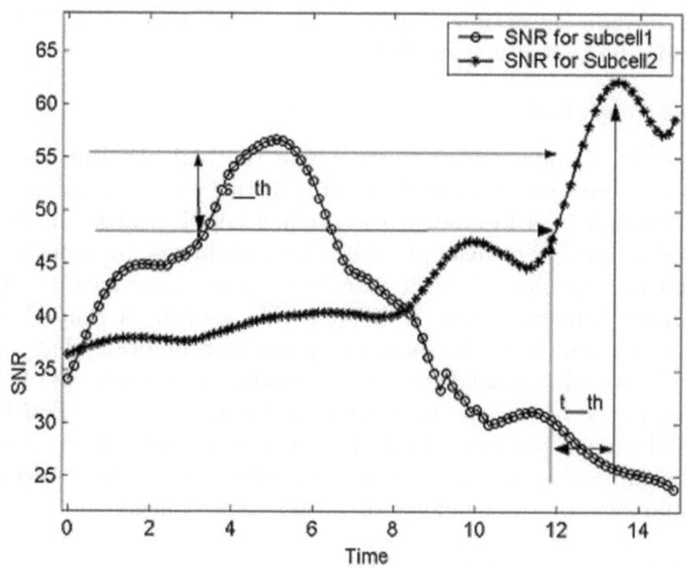

Fig. 2.4. SNR And Time Threshold For Handover Decision Making

C. Handover Execution

Handover execution is the last stage of the handover process. After the terminal has made the handover decision it will perform the handover execution stage. In this stage, the terminal attempts to synchronize to the handover target signal and to continue the reception of the currently received services without interruption. In order to validate whether the handover target signal is the correct one, the DVB-H signalling information contained in the TPS bits and the PSI/SI tables will be utilized. The handover execution stage is a frequency and transport stream synchronization stage. It consists of frequency synchronization using the TPS bits and PSI/SI tables and transport stream synchronization using the PSI/SI tables [105, 113, 73]. After the terminal has tuned to the correct frequency and transport stream, the terminal need also select the correct service (e.g. the consumed program) from the transport stream according to its obtained information within the received time slice.

2.2.2 Handover Challenges

Some challenges may exist in the handover process of DVB-H if the handover algorithm is not designed efficiently such as the *Ping Pong* effect, "fake signals" or tuning failure, excessive power consumption and packet loss. These are the challenges for DVB-H handover in general and the designed algorithms should cope with them. Also the network planning has big a effect on handover in DVB-H, as different network topologies decide whether a DVB-H handover is needed and how often is it needed.

A. Ping Pong effect

Because the signal strength fluctuates in the real physical environment the DVB-H receiver has the possibility of detecting strong signals from other cells even though it is located in the original cell, especially in the transmitter shadow areas. For example, when high buildings are blocking line of sight signal transmissions. In this case, the receiver may have the possibility of repeated handover between different cells, causing a *Ping Pong* effect [106]. Since frequent handover increases power consumption that is critical for battery powered handheld terminals, reducing the occurrence of the *Ping Pong* effect is one of the key research areas for handover in DVB-H. Since whether and when a handover should be performed is determined in the handover decision-making stage, the *Ping Pong* effect should be reduced to the minimum possible by the handover decision-making stage.

B. Tuning failure or "fake signals"

Tuning failure or "fake signals" refers to the situation where the DVB-H terminal makes supposedly the right handover to the target cell but actually handed over to another cell, which causes service interruption resulting from tuning failure. Handover in DVB-H utilises the PSI/SI tables within its received transport streams and information acquired from the TPS bits. The PSI/SI tables provide information to enable automatic configuration of the receiver to demultiplex and decode the various streams of programs within the multiplexed transport stream. In PSI/SI tables, there are different descriptors containing the signalling information for DVB-H. The information provided by the different descriptors is shown in Table 1.1 of Chapter 1. The terrestrial_delivery_system_descriptor and the frequency_list_descriptor in conjunction with the service_list_descriptor are the main parameters that are used in the simplest handover method introduced for the stationary and portable reception of DVB-T. Normally when a DVB-T terminal performs a handover, it will try to match the frequency of the strongest signal with the service id in its service_list_descriptor. If they match, it will perform the handover. However, if the terminal only uses these three parameters to perform handover tuning failure may result because these three descriptors are not enough to provide the match between exactly one service and one frequency.

The cell identification information, i.e. cell_id, is also signalled within the TPS bits of each received signal, which is critical for the receiver to discover the correct service. In other words, a "fake signal" is a signal that has the same frequency and cell_id as the targeted signal but which actually is from a different network and hence it is very unlikely that the receiver is able to receive currently consumed IP streams from it. Such a situation may occur for example when a cell of another network is using the same cell_id and frequency as the cell that the receiver aims to hand over to.

Tuning failure or "fake signals" is illustrated in Fig. 2.5. In Fig. 2.5 the two different DVB core networks are connected based on the two network operators agreement if they are operated by two different network operators.

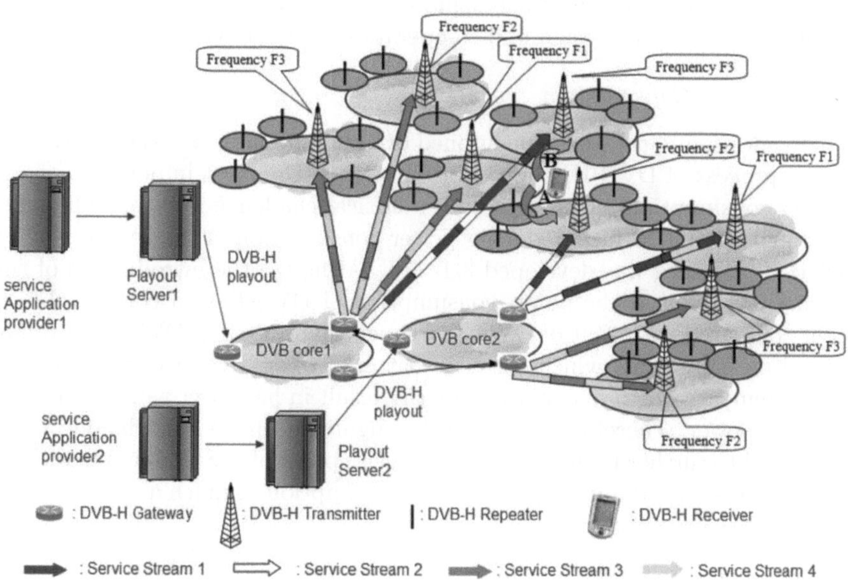

Fig. 2.5. Tuning Failure or Fake Signals

When the terminal moves from one cell to another as in case A shown in Fig. 2.5, it will detect a sufficient signal strength to trigger the handover process from the signals of frequency F2. As frequency F2 is matched with the desired service, the DVB-H terminal will perform handover from frequency F1 to frequency F2. However, after the terminal has handed over to F2, the user using the terminal will encounter a service interruption that is defined as tuning failure from the user's point of view. In case B, shown in Fig. 2.5, the DVB-H terminal will detect sufficient signal strength to trigger handover process from the signals of frequency F3. As frequency F3 is matched with the service the terminal is supposed to receive, the DVB-H terminal will perform handover from frequency F1 to frequency F3. Only after handing over to F3 does the user of the DVB-H terminal realize that there is a tuning failure. The difference between case A and case B is that case A happens when the terminal tries to synchronize with a frequency from another DVB core network where there are at least two different services using the same frequency. However, case B happens when the terminal tries to synchronize with the frequency from the same DVB core network where there are at least two different services using the same frequency. The two different services can be received on the same frequency because the two same frequencies in the network are far enough away from each other so that they do not interfere with each other even if they provide two different services. The signals causing tuning failure are also called "fake signals".

The tuning failure or "fake signals" described above can be avoided by appropriate network design and co-operation between network operators.

C. Power Consumption

In addition to the above-mentioned two main possible problems in the handover process of DVB-H, power consumption is the most important concern. Power consumption has always been a critical challenge for mobile handheld terminals [5, 6]. In fact, reducing power consumption is the reason why the DVB-H standard was developed [94, 110]. Although the introduction of time slicing has reduced the power consumption of DVB-H to a considerable extent compared with that of DVB-T, frequent handovers in DVB-H need more signal measurements which consume battery power. Even frequent handover measurements which may not necessarily result in handover will also consume battery power. Therefore, the handover algorithm in DVB-H should be fully exploited to further reduce the power consumption of the terminal in different stages and to avoid unnecessary power consumption as much as possible when handover is present.

D. Packet Loss

DVB-H is a unidirectional broadcasting network. If some packets are lost during the handover process there will be no retransmission of the lost packets. Packet loss will most probably happen when the terminal tries to synchronize to the target frequency and transport stream in the handover process. Delay and jitter are very common in the IP networks that are the service-feeding networks of DVB-H. Because of unidirectional nature of DVB-H no retransmission is possible except another uplink channel is utilized (e.g. UMTS). Since even a single lost packet will have a disastrous effect for some IP Datacast services in DVB-H (e.g. file downloading), strict synchronization techniques must be used in the synchronization of the time sliced services of DVB-H. Basically, packet loss is a fundamental issue in DVB-H that needs to be solved in any practical handover scheme for DVB-H. Further discussion of this issue can be found in [105, 113].

2.3 Designing a Better Handover Algorithm for DVB-H

The handover in DVB-H networks is the subject of on-going research and different approaches for designing handover algorithms by utilising mechanisms defined in [67, 68, 93] are developed all the time. In this section, some key points are presented as criteria for designing an efficient handover algorithm in DVB-H networks.

A. Handover Decision-making Stage

One of the key aspects in designing an efficient handover algorithm for DVB-H is to exploit the possibilities of reducing battery power consumption. The handover decision-making stage is the handover phase where the battery power consumption reduction can be fully exploited. The main objective in the handover decision-making stage is to try to predict the handover moment to reduce the number of off burst time intervals that are used for handover measurement.

One thing need to be noted is that most of the power consumption in the DVB-H service consumption cycle comes from the loading and playing of the media players when audio and video are involved. Although the front-end radio reception consumes less power compared with the media playing, it still makes difference regarding to the overall battery power consumption. Thus it should be considered in the handover design.

B. Complexity and Compatibility

The design of a handover algorithm for DVB-H should not conflict with the already consolidated DVB-H standards and the complexity of the handover algorithm should be fully exploited to ease the difficulty imposed on the receiver design.

C. Utilization of Additional Signalling Information

Additional signalling information should always be fully exploited by the handover algorithm. Handover in dedicated DVB-H networks has the characteristic of being passive where only the unidirectional transmission from the network to the terminal is possible. If additional signalling information is available, it should be used to help the handover process. Take the converged terminal as an example; converged DVB-H and GPRS/UMTS terminals have the advantage of having an interactive uplink channel. In this case, the uplink channel can be utilized to aid the handover process and this UMTS aided handover is a kind of active handover. The network parameters transmitted from transmitters and repeaters can also be fully utilized by the passive DVB-H receivers to aid the handover process.

D. Additional Equipment Cost

DVB-H terminals should be affordable for consumers. An additional attachment such as a GPS receiver can improve handover efficiency by predicting and checking the right handover moment. However, such an additional attachment will also increase the terminal price. From the economical aspect, the DVB-H handover algorithms should focus on utilizing existed signalling information that is available in the DVB-H standard to avoid the extra cost of introducing new network equipment (such as expensive repeaters) or terminal attachments (such as GPS receivers) if these extra equipments are used for handover purposes only.

In addition, the handover challenges presented in section 2.2.2 should be carefully considered when a handover algorithm is designed. The challenges such as the Ping Pong effect, "fake signals" and power consumption are sometimes related each other. For example, because more Ping Pong effect means more power consumption, when the Ping Pong effect is addressed, the power consumption is usually reduced as well.

Different handover algorithms have different characteristics. It is sometimes difficult for one algorithm to be used for all situations, for example, a handover algorithm utilizing UMTS interaction channels will not work when the UMTS network does not exist. While the above-presented evaluation criteria should be considered in designing an efficient handover algorithm for DVB-H, the individual application situation must also be taken into account. Designing an efficient algorithm usually implies a trade-off between power consumption, signalling information and additional equipment cost under the condition that the complexity and compatibility problems are considered.

Problems

2.1. What is IPDC?

2.2. What benefits does IPDC bring to DVB-H?

2.3. In which kind of DVB-H networks is handover required?

2.4. What are physical handover and service handover in DVB-H?

2.5. What are the different handover stages in DVB-H?

2.6. What are the handover challenges in DVB-H?

2.7. What are the issues that we should consider in order to design a better handover algorithm in DVB-H?

3

Survey of Handover Research in DVB-H

Handover in DVB-H is a novel issue. However, much research work has already been reported in this area. In this chapter, an attempt is made to survey as far as possible the work that has been reported in the DVB-H handover research.

3.1 Instantaneous RSSI Based Handover

An instantaneous Received Signal Strength Indication (RSSI) value based handover scheme was proposed in [104]. This handover scheme is the first for DVB-H published in the literature. This scheme uses the off burst time to measure the RSSI value. After comparing the current RSSI value with that of adjacent cells, it hands over to the cell with the strongest RSSI value. The handover stages for this handover algorithm are shown in Fig. 3.1 and Fig. 3.2.

The measurement and initialization stage shown in Fig. 3.1 is a fundamental stage that is conducted when the terminal just powers on. In this stage, the terminal scans the signals within the DVB-H frequency range (e.g. 470 - 890 MHz). If the terminal is synchronized with one frequency, it will search the NIT table to find the service that is matched with the synchronized frequency. Then it begins to store this service and frequency information in its memory. This procedure is repeated for all the available frequencies in the frequency range.

Fig. 3.2 shows the handover decision-making and execution stage. In Fig. 3.2 the terminal first measures the current signal strength in the off burst time of the time slicing mode. When the predefined RSSI degradation threshold is reached, it begins to measure the available signals in the adjacent cells. Then it synchronizes with the strongest signal with the biggest RSSI value. After synchronization to the new signal, it double-checks the handover accuracy using the cell_id information stored in its memory. When the handover to the target cell is assured, all the relevant information in the NIT tables are updated and the terminal repeats the process that may lead to another handover decision-making and execution stage.

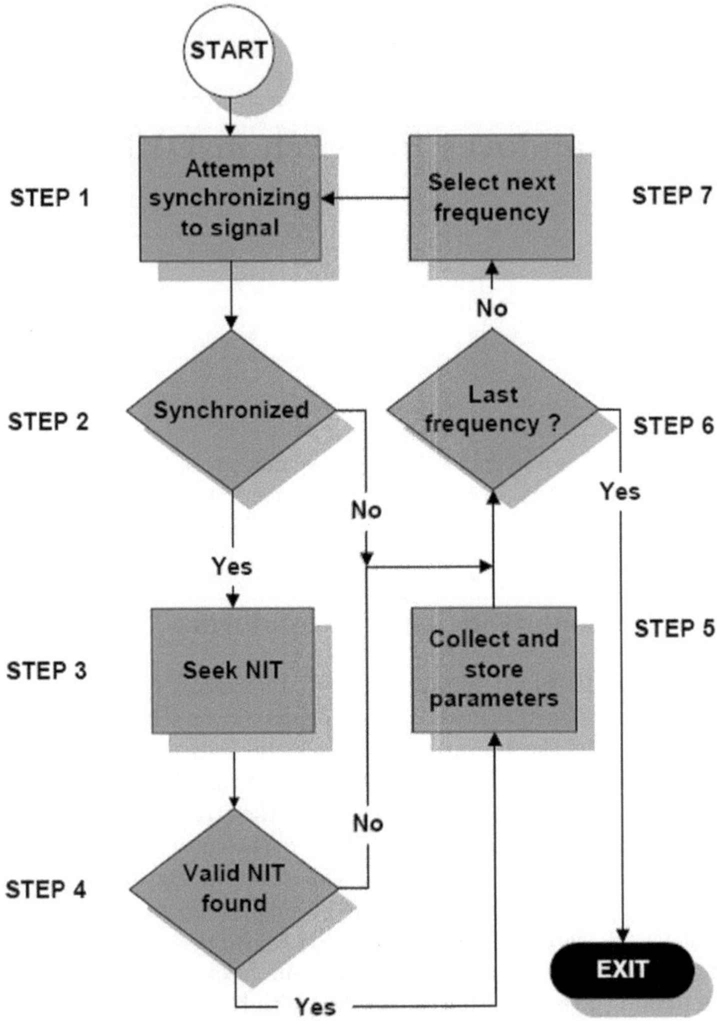

Fig. 3.1. Measurement and initialization Stage of the Handover Algorithm according to [104]

Since the RSSI value can vary due to multipath, interference or other environmental effects it may not give a true indication of the communication performance or the range and mistakenly measuring the RSSI value would result in the *Ping Pong* effect in handover measurement consuming power unnecessarily. The RSSI value could be measured many off burst times with the RSSI value being measured at least once every off burst time in the worst case. This scheme cannot eliminate effectively the possibility of receiving "fake signals", either [104]. Constant measuring of the adjacent cells signal level

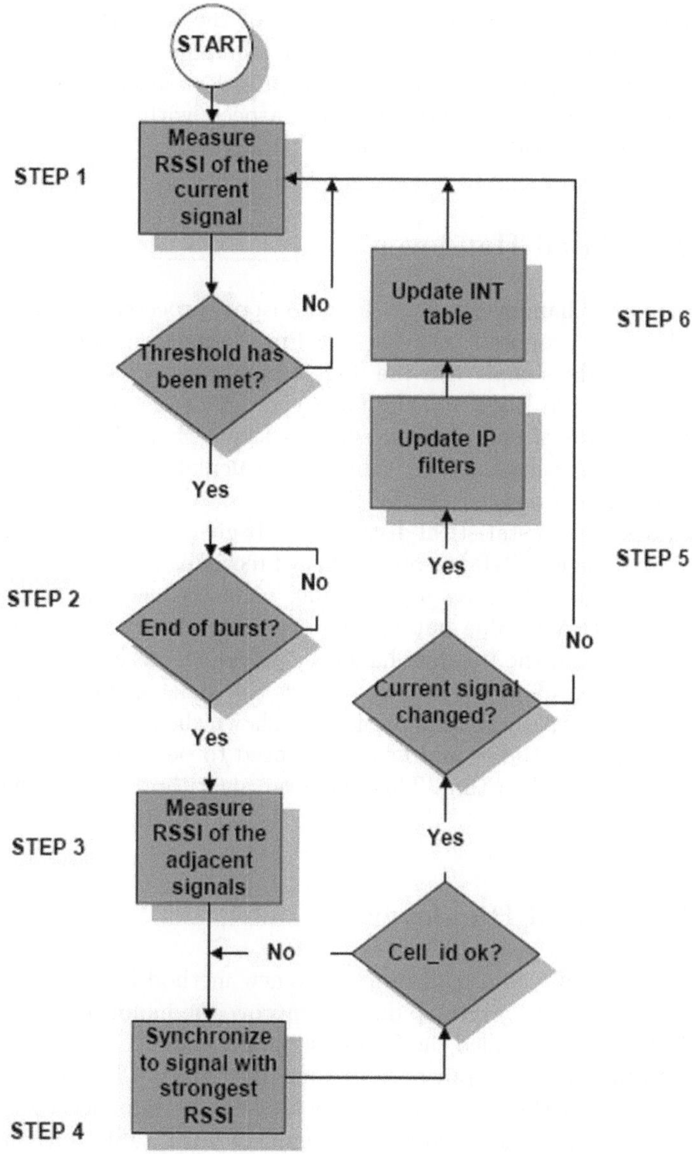

Fig. 3.2. Decision Making & Execution Stages of the Handover Algorithm according to [104]

without any handover prediction leads to more battery power consumption and receiving "fake signals" leads to degraded quality of service. In order to overcome these shortcomings a better handover prediction algorithm has to be added to enhance this RSSI based algorithm.

Hamara presented an enhanced version of the algorithm of [104] in his thesis [42] where in addition to using the RSSI value as a handover criterion currently consumed services and bit error rate were taken into account. This gives an extensive analysis of the handover aspects within DVB-H in the light of the standard solutions at the time.

3.2 SNR Based Handover

In [106] another handover scheme based on post-processing of the measured SNR value was proposed to avoid the Ping Pong effect and to get rid of the received "fake signals". In the SNR based handover scheme, the SNR is calculated from the RSSI and the noise characteristics and provides a more accurate estimate of the received effective signal than the RSSI. The main idea of post-processing the SNR values is to calculate the CDFs (Cumulative Distribution Functions) of all the SNR values. A Cumulative Distribution Function describes a statistical distribution. It gives at each possible outcome of the received signal SNR the probability of receiving that outcome or a lower one. Because the CDF gives a probability value, its value depends not only on the current SNR value but also on the SNR history of the signal. This not only eliminates the frequent handover phenomenon seen in instantaneous RSSI value based handover but also avoids the "fake signals" caused by frequency confusion. Although simulation has shown the feasibility of this simple algorithm, further studies and field trials need to be done to investigate the limitations of this algorithm. This handover algorithm is described in detail in Chapter 7.

3.3 CDT Based Handover

Vare, Hamara and Kallio [111] proposed a new method for signalling cell coverage areas by means of bitmap data to improve the handover performance in DVB-H. A new table called the Cell Description Table (CDT) is proposed for the PSI/SI tables in the transport streams. By using a CDT up to 256 different signal levels within the cell coverage area can be signalled to the receiver to inform it of the cell coverage. This kind of CDT table tells the receiver where it is located within the cell according to the different signal strengths in different locations within the transmitter coverage. The terminal can make better handover decisions from the information about the cell coverage area to reduce the *Ping Pong* effect and "fake signals". However, in this proposed handover scheme more bandwidth and receiver memory consumption will be needed to support the CDT information process. In addition the DVB-H handheld receiver must have Global Positioning System (GPS) support which will be an additional cost to the customer. This kind of cost cannot be neglected especially in the early DVB-H roll out stage when potential customers are still not fully convinced of the benefits of DVB-H services. Transferring the cost from the terminal side to the network side is an alternative solution if the network can provide the same location information to the terminal as a GPS receiver does.

One alternative is to use the repeater aided handover algorithm described in the following section.

3.4 Repeater Aided Handover

Repeater Aided soft handover was proposed in [25]. In the repeater Aided soft handover scheme, there are intelligent repeaters located in the cell border area. These repeaters not only enhance the signal quality in the cell border area, but also provide special handover signaling information to the repeaters. The basic idea is that the terminals will receive unique signalling information from the repeaters when the terminals move to the border area. In this way, the terminal can know where it is located depending on the different signalling information it receives from the repeaters. Using this special signaling information, the terminal can trigger handover process. By doing this, the terminals do not need to keep measuring signals for handover when they are far from the cell borders, thus save batter power that would otherwise be used for signal measurement.

The details of the repeater aided handover algorithm are described in Chapter 8 and Chapter 9 of this book.

3.5 Fast Scattered Pilot Synchronization Based Handover

Schwoerer [105, 113] and Vesma [105] target power consumption reduction by utilizing novel synchronization techniques in the handover execution stage of the handover process. The handover execution stage is equal to the signal synchronization and the on-time in the time slicing mode consists of both the synchronization time and the burst data duration time as shown in Fig. 3.3, the main idea of [105] and [113] is to try to minimize the synchronization time to further reduce the power consumption. The main exploitable synchronization time in the synchronization stage is the TPS synchronization as shown in Fig. 3.4 [105].

The synchronization of DVB-H signals is actually the synchronization on OFDM receivers, because all the DVB-H signals are modulated using the OFDM modulation scheme. The synchronization for OFDM receivers can be done either before or after the demodulating via Fast Fourier Transform (FFT), which is called Pre-FFT synchronization and Post-FFT synchronization [131]. For DVB-H signals synchronization, the TPS synchronization is the main synchronization stage that is directly related to the DVB-H physical layer specifications [70]. The channel estimation stage is the last stage of the DVB-H synchronization phase. The channel estimation stage must estimate both the channel and any residual phase errors [32]. It should be specifically designed for the channel estimation in the mobile environment because DVB-H is mainly designed for transmission to mobile receivers.

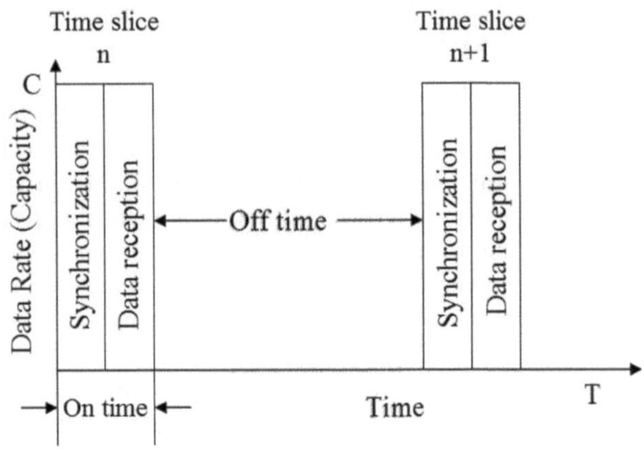

Fig. 3.3. Position of the Synchronization Duration in the Time Slicing Mode

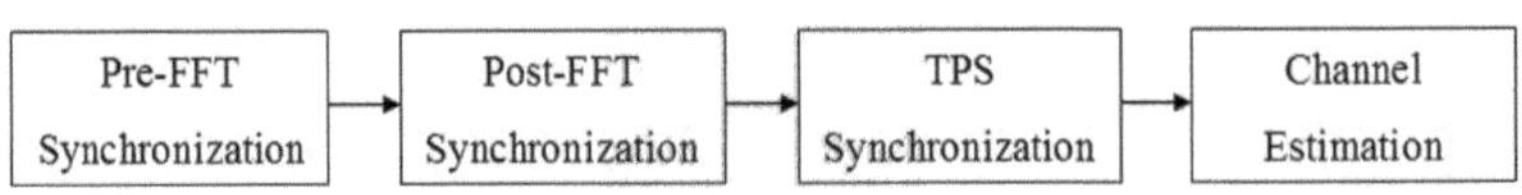

Fig. 3.4. TPS Synchronization in the Synchronization Stages According to [105]

Schwoerer and Vesma [105] also proposed a new synchronization technique called correlation-based "Fast Scattered Pilot Synchronization" for DVB-H receivers to substitute the conventional TPS based OFDM frame synchronization for finding the position of Scattered Pilots within an OFDM frame in the handover execution stage. The Scattered Pilots are training symbols that form a periodic pattern with specific period in time and in frequency. The Scattered Pilots are transmitted at a boosted power level to facilitate the synchronization of the OFDM frames. The correlation-based "Fast Scattered Pilot Synchronization" algorithm exploits the temporally repetitive structure of the scattered pilots and Schwoerer and Vesma [105] showed using mathematical analysis that the synchronization time (until channel estimation) could be cut by 84% by using the new technique. Reducing the synchronization time means reducing the power consumption because the terminal has to be powered on to make the synchronization. Therefore, "Fast Scattered Pilot Synchronization" can reduce the power consumption in the DVB-H handover execution stage. Schwoerer [113] proposed another purely power-based "Fast Scattered Pilot Synchronization" method. It uses the fact that scattered pilots are amplitude-boosted by 4/3 to find the current Scattered Pilot Raster Position (SRPR) [70]. It is shown in [113] that 89% of the synchronization time can be saved using power-based "Fast Scattered Pilot Synchronization".

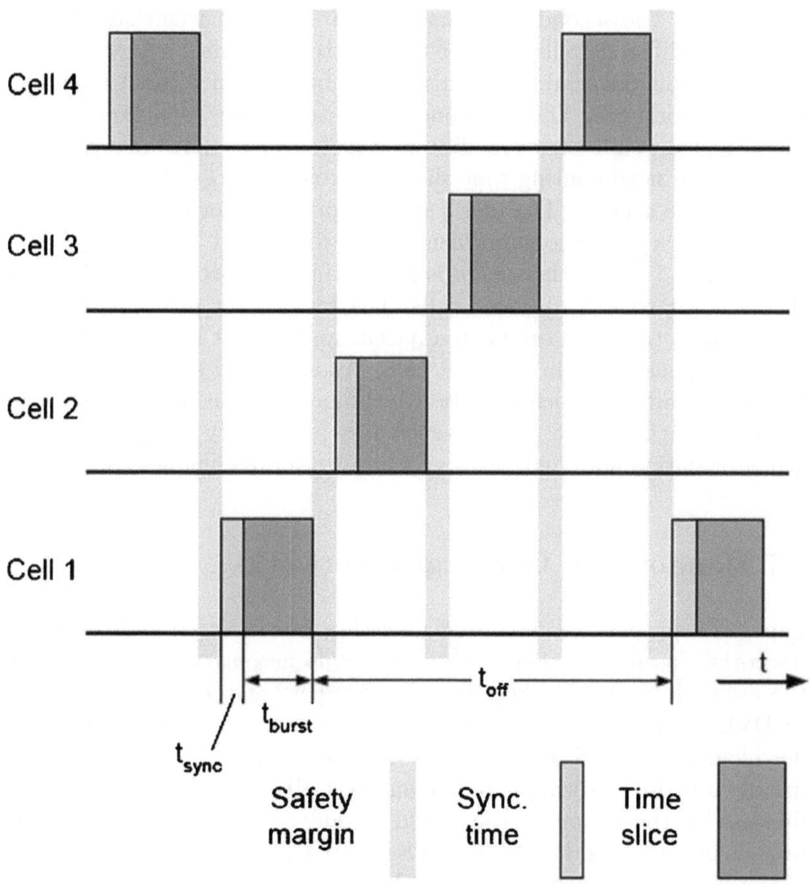

Fig. 3.5. Phase Shifting As a Four-colour Problem According to [102]

3.6 Phase Shifting Based Handover

In [102] May is focusing on the handover execution stage of the handover process in DVB-H. May proposed a technology called "phase shifting" to synchronize the signals of adjacent cells in IP Datacast over DVB-H networks in order to ensure loss-free handovers because the IP network delay and jitter maybe different for different cells, when the terminal moves from one cell to another, synchronization techniques must be used to ensure that there is no packet loss caused by a time sliced burst overlap when the next time sliced burst arrives. There are three different possibilities for the types of synchronization. The first is no synchronization that of course will cause considerable packet loss. The second is in-phase synchronization where all the transport streams in different cells must be transmitted in perfect synchronization that is at the same universal time. This cannot be ensured without a buffer system in the network side. The third one is "phase shifting" synchronization where there is a time shift between adjacent cells to ensure that there is enough time between the neighbouring time slices to avoid the possible packet loss caused by time slice overlap. The phase shifting principal for handover between any two cells as a four-colour problem [10] is illustrated in Fig. 3.5 according to [102]. Fig. 3.5 shows that there is no overlapping between the time slices of the adjacent cells of any four cells. The four colour problem solution never allows twice the same colour in adjacent nodes, that is, it never allows twice the same phase shift in adjacent cells. Analysis and simulation showed that the phase shifting synchronization techniques can achieve much better performance with respect to the packet loss probability compared with the no synchronization and "in phase" synchronization techniques [102].

3.7 Handover in Converged Networks

Paper [27] proposed a handover algorithm in converged DVB-H and UMTS networks. In such converged networks, it is necessary to optimally relocate the available radio resources (e.g. bandwidths), i.e. allocate the users either to DVB-H and UMTS when the users are receiving the same streaming or download services. Stochastic trees are open-ended Markov chains which are usually used for medical prediction analysis [19]. In [27] the stochastic trees are first used in the communications field to analyze the intersystem soft handover process in converged DVB-H/UMTS networks. The investigation and research of the handover in converged DVB-H and UMTS networks presented in [27] are thought to be one of the first in the literature. Details of this kind of handover algorithm are described in Chapter 10 of this book.

3.8 Handover Proposed By DVB Project

The DVB Project technical reports [69] and [68] proposed a simple handover algorithm for handover in DVB-H based on the handover algorithm in DVB-T. The basic idea is for the DVB-H terminal to use the terrestrial_delivery_system_descriptor, the frequency_list descriptor, the original_network_id and the transport_stream_id together as a pair along with the service_list_descriptor to decide which frequency and transport stream the receiver should switch to in the handover process. Several methods to reduce the risk of tuning failures or "fake signals" are also presented. The first one proposed in [69] is called "local SI insertion" that makes each cell a separate network by individual Service Information (SI) insertion. In this case, there will be only one frequency per network by giving each network a separate frequency indicated in the SI table. The second method utilizes the cell identifier so the terminal can know which cell it has entered. In this case, the terminal can determine and check the cell ID of a signal from its TPS bits to see if it is in its cell ID list of interest after checking the frequency thus reducing the tuning failure. The third method uses location data from GPS receivers to aid the handover so the terminal can determine the destination cell reducing tuning failure. The last method in [69] uses two front-ends including a second demultiplexer. In this case, the tuning of different frequencies can be done in parallel and the target cell frequency can be validated in advance so that the risk of tuning failure can be completely eliminated. However, this method is largely dependent on the size of terminal. Because two front-ends will certainly occupy more space in the receiver. From another aspect, two-front-ends receiver also consumes more power compared with single-front-end receiver.

3.9 Research Projects Related to DVB-H Handover

A myriad of projects on DVB-H were finished or are being carried out around the world. Each of the projects focuses on one specific topic regarding to the DVB-H handover. Two examples are given here:

3.9.1 IST INSTINCT

INSTINCT was a European IST IP (Integrated Project) Project on IP-based Networks, Services and Terminals for Converging Systems, namely the convergence of Broadcast networks (DVB-T/H) and mobile Telecommunications networks (GPRS/UMTS) [137]. INSTINCT started on 1 January 2004 and lasted for two years. Handover issues in DVB-H and in converged DVB-H/UMTS networks are researched by Brunel University, France Telecom R&D, and TDF in the first phase of the INSTINCT project [137, 109].

3.9.2 IST MING-T

MING-T is a European IST STREP (Specific Targeted Research Project) Project on Multistandard integrated network convergence for global mobile and broadcast technologies [140]. MING-T is targeting on hybrid broadcast networks where vertical handover between different broadcast networks are required. As a result, MING-T helps the interoperation between different broadcast standards, particularly DVB-H and one of the Chinese broadcast standards DMB-T.

3.10 Conclusion

The DVB-H standard leaves much space for the handover algorithm design and development. With the rollout of DVB-H commercial services in each country, designing and choosing of the appropriate DVB-H handover algorithms begin to come into the schedule of DVB-H rollout plans.

Problems

3.1. Why is RSSI based handover liable to Ping Pong effect?

3.2. What are the hardware requirements for the terminal to implement CDT based handover?

3.3. What are the functions of the Scattered Pilot within an OFDM frame?

3.4. How is the Phase Shifting realized in the Phase Shifting based handover?

3.5. What are the handover algorithms proposed by DVB Project?

4

DVB-H Signalling Information

4.1 Introduction

The signalling information in DVB-H is mainly used by the receiver for service discovery. It includes PSI/SI tables in the data link layer, the TPS bits in the physical layer, the ESG and EPG in the application layer.

Even though DVB-H was designed to be backwards compatible with DVB-T, there are differences between the signaling in DVB-T and the signaling in DVB-H.

Regardless of the fact that DVB-T and DVB-H share some common PSI/SI tables, such as the NIT, PAT, PMT, INT and TDT (Time and Date Table), the DVB-H receiver does not need to support the SDT (Service Description Table) and EIT (Event Information Table). In [87], the SDT and EIT tables are considered mandatory for an IPDC over DVB-H network but optional for the receiver. Also the linkage modes to enable support for handover to associated services, as defined in [69], are not supported in [87] and hence are not discussed within this chapter.

4.2 PSI/SI Tables

The Program Specific Information (PSI) was first defined within ISO/IEC 13818-1 [71], in order to make the Integrated Receiver Decoder (IRD) automatically configure itself for the selected service transmitted using MPEG-2 transport streams. It contains the following four types of information tables [72]:

1. Program Association Table (PAT):
   - for each service in the multiplex, the PAT indicates the location (the Packet Identifier (PID) values of the Transport Stream (TS) packets) of the corresponding Program Map Table (PMT). It also gives the location of the Network Information Table (NIT).

2. Conditional Access Table (CAT):
   - the CAT provides information on the CA systems used in the multiplex; the information is private (not defined within the present document) and dependent on the CA system, but includes the location of the EMM stream, when applicable.
3. Program Map Table (PMT):
   - the PMT identifies and indicates the locations of the streams that make up each service, and the location of the Program Clock Reference fields for a service.
4. Transport Stream Description Table (TSDT)
   - TSDT provides information about the entire Transport Stream, for example the type of target receiver (DVB, ATSC) or the kind of application (e.g. satellite contribution link). All descriptors carried within the table apply to the entire Transport Stream.

The DVB Project further extended the PSI defined in ISO/IEC 13818-1 and named it Service Information (SI). Therefore, the SI information tables are usually called PSI/SI tables. The DVB Project defined the data format for NIT table and make it mandatory for DVB transport streams. The DVB Project also defines ten other PSI/SI tables. Thus the fourteen information tables contained in DVB are:

1. Program Association Table (PAT):
2. Conditional Access Table (CAT):
3. Program Map Table (PMT):
4. Transport Stream Description Table (TSDT)
5. Network Information Table (NIT)
   - the location of the NIT is defined by DVB Project in compliance with ISO/IEC 13818-1 [71] specification, but the data format is outside the scope of ISO/IEC 13818-1 [71]. The NIT is intended to provide information about the physical network.
6. Bouquet Association Table (BAT)
   - the BAT provides information regarding bouquets. As well as giving the name of the bouquet, it provides a list of services for each bouquet.
7. Service Description Table (SDT)
   - the SDT contains data describing the services in the system e.g. names of services, the service provider, etc.
8. Event Information Table (EIT)
   - the EIT contains data concerning events or programmes such as event name, start time, duration, etc.;

- the use of different descriptors allows the transmission of different kinds of event information e.g. for different service types.

9. Running Status Table (RST)
   - the RST gives the status of an event (running/not running). The RST updates this information and allows timely automatic switching to events.
10. Time and Date Table (TDT)
    - the TDT gives information relating to the present time and date. This information is given in a separate table due to the frequent updating of this information.
11. Time Offset Table (TOT)
    - the TOT gives information relating to the present time and date and local time offset. This information is given in a separate table due to the frequent updating of the time information.
12. Stuffing Table (ST)
    - the ST is used to invalidate existing sections, for example at delivery system boundaries.
13. Selection Information Table (SIT)
    - the SIT is used only in "partial" (i.e. recorded) bitstreams. It carries a summary of the SI information required to describe the streams in the partial bitstream.
14. Discontinuity Information Table (DIT)
    - the DIT is used only in "partial" (i.e. recorded) bitstreams. It is inserted where the SI information in the partial bitstream may be discontinuous.

Update Notification Table (UNT)

- The main function of the UNT table is to describe the availability and location of the System Software Update services. UNT makes the IPDC DVB-H Receiver know where, when and how the System Software Update services can be found.

System Software Update (SSU)

DVB-H adds another PSI/SI table to the DVB PSI/SI tables set. It is:

1. IP/MAC Notification Table (INT)
   - The main function of the INT table is to make the DVB system more suitable for IP stream based signalling. The INT describes the availability and location of IP streams within a DVB MPEG2 transport stream. Thus the INT tables are important in the DVB-H handover execution stage for the receiver to synchronize to the targeted IP streams.

Since the PSI/SI tables are transmitted within the transport stream in the DVB-H networks, in order to discover and consume the service, the DVB-H receiver must implement the PSI/SI tables according to [87] during the receiver design.

4.3 TPS Information

The Transmission Parameters Signalling (TPS) bits are used to signal parameters related to the transmission scheme, i.e. the parameters related to channel coding and modulation [87]. The TPS carriers convey information on:

- constellation including the α value of the QAM constellation pattern (the α value defines the modulation based on the cloud spacing of a generalized QAM constellation. It allows specification of uniform and non-uniform modulation schemes, covering QPSK, 16-QAM, and 64-QAM);
- hierarchy and interleaving information;
- guard interval;
- inner code rates;
- transmission mode;
- frame number in a super-frame;
- cell identification;
- time-slicing indicator;
- MPE-FEC indicator

The TPS is defined over 68 consecutive OFDM symbols, referred to as one OFDM frame. Four consecutive frames correspond to one OFDM super-frame. Each OFDM symbol conveys one TPS bit. Each TPS block (corresponding to one OFDM frame) contains 68 bits, defined as follows:

- 1 initialization bit;
- 16 synchronization bits;
- 37 information bits;
- 14 redundancy bits for error protection.

The eight TPS bits s_40 to s_47 are used to identify the cell to which the DVB signal belongs. These bits contain the cell_id. In the handover decision-making stage, the receiver will rely on the cell_id to handover to the target signal.

Note that cell identification is 16 bits. Therefore when carried in TPS bits, the Most Significant Byte (LSB) and Least Significant Byte (MSB) of the cell_id are interleaved.

4.4 Electronic Service Guide

Electronic Service Guide (ESG) contains information about the services available. Through the user interface to the ESG, the user can select the services and items that he/she is interested in and finds pre-stored service links on the terminal.

The ESG operation takes place after the DVB-H receiver has synchronized to a DVB-H transport stream. The ESG operation contains three stages: ESG bootstrap, ESG acquisition and ESG update.

- ESG bootstrap
  - The ESG bootstrap stage is the service discovery stage, where the DVB-H terminal looks for the ESG bootstrap session IP address in the received PSI/SI tables. The ESG bootstrap process provides the different available ESGs relating to different DVB-H platforms. In other words, the ESG bootstraps can provide different ESGs from different DVB-H operators, thus providing a natural way for operator differentiation.
- ESG acquisition
  - The ESG acquisition stage refers to the process of the DVB-H terminal obtaining the different ESGs that are refereed to in the ESG bootstrap session. In another word, in the ESG acquisition stage the DVB-H terminal gets different ESGs according to the terminal design from the different DVB-H operators.
- ESG update
  - The ESG update stage refers to the process that the DVB-H terminal updates the ESG stored in itself either automatically after certain time or manually by the users. The ESG update process keeps the DVB-H terminal always updated with the current available services.

4.4.1 Service Description Protocol

The Session description Protocol (SDP) [64] file is part of the ESG and it contains information that the terminal needs in order to be able to receive and consume the content of a service, namely audio/video service. Every session description file relates to a service or a schedule event of a service. A session description file contains application configuration information such as the information to receive the service (addresses, ports, formats and so on).

There are two ways of transmitting SDP files within the ESG. Inline and Out of Band. Inline refers to that the SDP file is contained in the SessionDescriptionType element. Out of Band refers to that the SDP file is referenced in the SessionDescriptionType element.

Each time the service parameters are changed, for example the encoding bit rates are changed, the SDP file related to the service should be updated in the ESG server. In this way, the terminal can get updated in time for the service change so as to function properly.

4.5 Electronic Program Guide

Electronic Program Guide (EPG) conveys on the available services to the users in a more visible way. It is the direct interface between the user and the available services. A typical usage is that the EPG delivers the information of the current available TV channels and the schedule of the future TV programs. EPG is written using XML on the ESG server and it must be continuously updated as the TV programs are preceding.

4.6 Analysis of DVB-H Signalling

As DVB-H is utilizing MPEG-2 transport stream for transmission just as DVB-T, DVB-H signals can also be received by a normal DVB-T receiver. The usual way for the DVB-H signalling analysis is: after the DVB-H transport streams are captured by the DVB-T receiver, the MPEG-2 transport stream can be decoded into IP streams and finally the information about the PSI/SI tables and TPS bits can be obtained. Higher level application softwares can be used for further analysis. Such signalling information obtained on the receiver side can be used as a comparison with that from the transmission of the head-end (transmission) side. In case an error occurs, such comparison is very useful for the problem tracking and debugging. Different commercial hardwares and softwares are already available for the analysis of DVB-H signalling information.

4.7 Conclusions

Within the DVB-H protocol stack, different layers have different signalling information. As the DVB-H standard consists mainly of the signalling of the physical layer, data link layer and the application layer, only the related signalling information regarding to the three layers are presented. They are PSI/SI (data link layer), TPS (physical layer), ESG and EPG (application layer).

Problems

4.1. What are the main DVB-H signalling information and in which protocol stack layers are they located?

4.2. What is SDP file used for?

4.3. What are the different ways to transmit a SDP file within ESG?

5

Electronic Service Guide

5.1 Introduction

ESG is the service discovery tool both for the consumers and for the client applications on the mobile terminal. The ESG provides the consumers with rich, up-to-date information about the services. ESG also serves the mobile terminal middleware with signaling data to enable service lookup from the DVB-H stream and playback with the correct client software and codecs.

There are two groups of ESG standards. One is from the DVB group, called IPDC ESG [77]. The current IPDC ESG version is 1.0 with version 2.0 in the development stage. The other group of ESG is from the OMA group, called OMA BCAST ESG [78]. The current OMA BCAST ESG version is 1.0, which is still in the finalizing stage at the writing of this book while the BMCO forum has already tested and declared a preliminary commercial version of OMA BCAST ESG.

The term "Service Guide" and "ESG" in this chapter are identical regarding to the meaning.

5.2 IPDC ESG

5.2.1 IPDC ESG Layers

The IPDC ESG specification covers the description of the data model, the representation, the encapsulation and the transport, as shown in Fig. 5.1.

The IPDC ESG data model defines the ESG fragments using XML. The ESG encapsulation is divided into three parts: ESG Container (used to facilitate the processing and transmission of ESG information of considerable size), ESG Fragment Management Information (facilitates the management of the ESG fragment without looking at the contents of the fragments) and ESG Data Repository (facilitates the fast random access to the content of the ESG Fragments). The ESG is transported using FLUTE protocol. Details are described in [77].

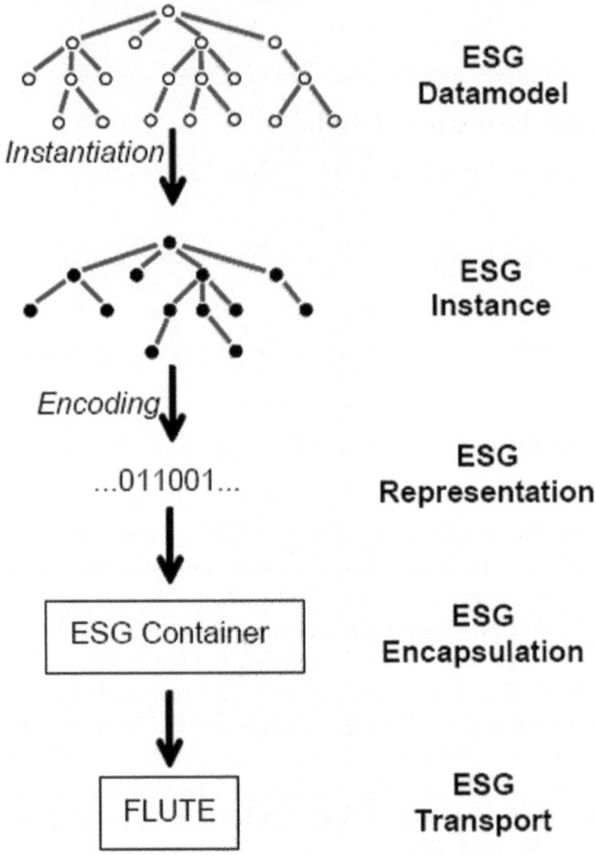

Fig. 5.1. IPDC ESG Specification Structure [77]

5.2.2 IPDC ESG Bootstrap Processing Flow

ESG operations take place after the DVB-H receiver has been started and the terminal is synchronized to a particular transport stream carrying IPDC services. Based on the ESG information rendered to a user through an ESG application, a specific service can be selected. The ESG also provides information which enables the terminal to connect to the related IP stream in the DVB-H transport stream. The IPDC ESG boostrap processing flow is as follows shown in Fig. 5.2.

Two descriptors are involved in the bootstrap process:

1. ESGProviderDiscovery Descriptor
2. ESGAccessDescriptor

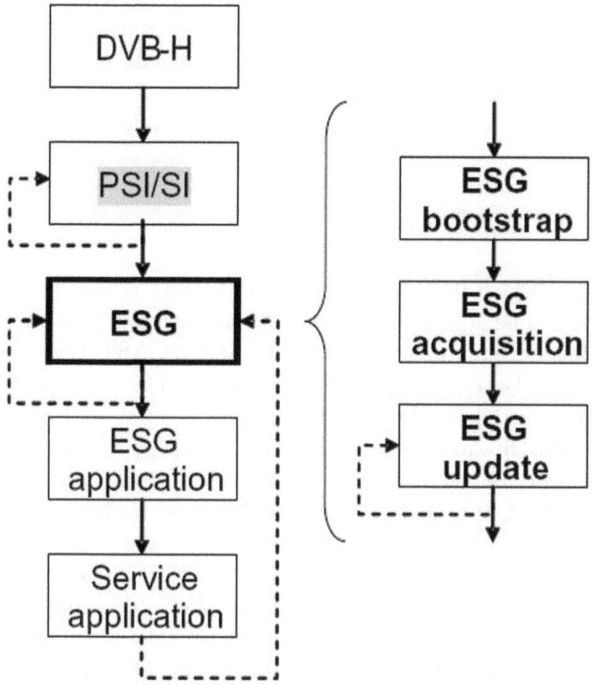

Fig. 5.2. IPDC ESG Boostrap Processing Flow [77]

Both descriptors are delivered through a FLUTE session with a well-known (registered) destination IP address and port (224.0.23.14 for IP Version 4 and FF0X:0:0:0:0:0:0:12D for IP Version 6 on port 9214 defined in section 9.2 of [77]). Furthermore, this session is the only one sent to that address and port, so that the terminal does not require any additional information e.g. Transport Session Identifier (TSI) of the session to start the bootstrap process [92].

The ESGProviderDiscovery includes a ProviderID, which is a unique identifier in the scope of the IP Platform associated with each described ESG-Provider. After the user has selected a particular ESG, the terminal uses the ProviderID to identify a particular ESG Entry within the ESGAccessDescriptor file, which contains information on how to acquire the ESG of a given provider. In this way, the operator separation is realized in the terminal by using the ESG.

5.2.3 DVB IPDC 1.0 and 2.0

The main difference between the DVB IPDC 1.0 ESG and the 2.0 ESG is that the DVB IPDC 2.0 includes a new feature: On-demand ESG through 3G (Unicast distribution as an option). DVB IPDC 2.0 also includes a notification feature which is similar to OMA BCAST ESG where the ESG can be used to delivery notification information to the terminals. Regarding the feature of interactivity, DVB IPDC 2.0 ESG also incorporates more interactivity features than the DVB IPDC 1.0 ESG. However, the DVB IPDC 2.0 is still in the development stage. New features could still be developed in the future.

5.3 OMA BCAST ESG

Just as DVB IPDC ESG, the OMA BCAST ESG is also located at the application layer. The lower radio bear layers are independent of the above application layers. This means that the OMA BCAST ESG can also utilize the DVB-H radio bear, which is called Broadcast Distribution System (BDS).

If DVB-H is used as a BDS, the DVB-IPDC bootstrap ESG is reused. And the ESGProviderDiscovery Descriptor and ESGAccessDescriptor are used to allow the discovery of the provider of the OMA BCAST ESG and the access to the OMA BCAST ESG, i.e. the ProviderID is used to distinguish the different ESG providers.

The ESGAccessDescriptor then links to the Service Guide Announcement Channel in the OMA ESG. Independent of the underlying BDS, the different Operators are also distinguished by using ID (bcast://operator Y.com or bcast://operator X.com) and BSMFilterCode (OPERATOR_X or OPERATOR_Y) and TransportObjectID (2 or 9) Id, BSMFilterCode and TransportObjectID are within SGDD. The ID and BSMFilterCode are listed below:

1. ID
   - Unique identifier of the SGDD within one specific SG
2. BSMFilterCode
   which contains the following attributes and elements:
   - type
   - serviceProviderCode
   - corporateCode
   - serviceProviderName
   - nonSmartCardCode
   - 3GPPNetworkCode
   - 3GPP2NetworkCode

There are two important concepts in OMA BCAST ESG. ServiceGuideDeliveryDescriptor (SGDD) and ServiceGuideDeliveryUnit (SGDU).

1. ServiceGuideDeliveryDescriptor (SGDD)
   - The ServiceGuideDeliveryDescriptor is transported in the Service Guide Announcement Channel, and informs the terminal the availability, metadata and grouping of the fragments of the Service Guide in the Service Guide discovery process. A SGDD allows quick identification of

the Service Guide fragments that are either cached in the terminal or being transmitted. For that reason, the SGDD is preferably repeated if distributed over broadcast channel. The SGDD also provides the grouping of related Service Guide fragments and thus is a means to determine the completeness of such group.
The ServiceGuideDeliveryDescriptor is especially useful if the terminal moves from one service coverage area to another. In this case, the ServiceGuideDeliveryDescriptor can be used to quickly check which of the Service Guide fragments that have been received in the previous service coverage area are still valid in the current service coverage area, and therefore don't have to be re-parsed and re-processed.
SGDD within the service guide declare the existence of and availability of service guide fragments. The SGDD allows the terminal to deduce which fragments are associated with which Mobile Broadcast Service Provider (through BSMFilterCodes).
Each BSMFilterCodes is corresponding to one Service Provider.

2. ServiceGuideDeliveryUnit (SGDU)
   - SGDU is the structure that the network uses to encapsulate fragment subsets for the transport layer. SGDU contains SDPfragment which is a String containing the actual SDP data, without termination character. SGDU can be sent plain or GZIP compressed.

5.3.1 Service Guide Discovery over Broadcast Channel

When the Service Guide is delivered using the broadcast channel the Service Guide Announcement Channel is thought as the starting point of the retrieval. Recall that the Service Guide Announcement Channel provides all the information the terminals need for retrieving the Service Guide. Therefore to discover the Service Guide the terminals basically need to locate the file delivery session carrying the Service Guide Announcement Channel. The access parameters of the FLUTE session representing the Service Guide Announcement Channel are called the entry point to a Service Guide on a Broadcast Channel. In one broadcast area there MAY exist multiple Service Guides and any number of these MAY be delivered simultaneously using the broadcast channel. In such a case, in principle, it is the responsibility of the underlying BDS to provide the signalling of the entry points of the Service Guides to the terminals. However, if such a signalling is not available or being used, the following parameters should be used as the entry point:

1. (OPTIONALLY) IP Source Address
2. Fixed Destination Multicast IPAddress: 224.0.23.165 for IPv4 or FF0X:0:0:0:0:0:0:132 for IPv6
3. Fixed Destination Port: 4090

The terminal should assume that:

1. there is at most one FLUTE session per entry point.
2. the value of the Transport Session Identifier (TSI) may be any valid value and the number of ALC/LCT channels in the FLUTE session for Service Guide announcement is fixed to 1.

If the underlying BDS supports specific signalling of the Service Guide entry points the terminal shall expect the BDS also to provide the specific signalling. The detailed guidelines for such signalling in specific Broadcast Distribution Systems are given in the BDS Adaptation Specifications (See [79], [80] and [81]).

The terminal shall support the initial Service Guide discovery over Broadcast Channel.

5.3.2 Service Guide Discovery over Interaction Channel

The entry point to a Service Guide on an Interaction Channel should be defined as the Uniform Resource Locator (URL) to a file containing Session Description or URL to a resource resolving to a Session Description which describes the file distribution session carrying Service Guide announcement information and possibly Service Guide. This file distribution session originates from Service Guide Generation Function and Service Guide Distribution Function. The entry point to a Service Guide on an Interaction Channel may be either fixed, or provisioned to the terminal (e.g. through BDS specific signalling), or provided out-of-band (e.g. through a public or private web site).

Within a single BDS, there may be different Service Guides generated for different service coverage areas, requiring a different entry point for each particular service coverage area. In this section, that how the device learns about the applicable URL will not be described.

The terminal with interaction channel should support the initial Service Guide discovery over Interaction Channel.

5.3.3 Service Guide Transmitted over Interaction Channel

The service guide discovery mechanisms that are specified in this section relate to the discovery of a Service Guide that is to be distributed over Interaction Channel. The Terminal needs to get some discovery information, and sends the request to acquire Service Guide.

The entry point to Service Guide acquisition over Interaction Channel should be a URL which indicates the location of Service Guide. Example of such URL is http://provider.com/serviceguide. This is the address that the Terminal accesses in order to get Service Guide data over Interaction Channel. There are several possible ways a terminal can get the entry point information. The Terminal should support the following two means:

1. the entry point information is provided using the 'AlternativeAccessURL' element of SGDD;
2. the entry point information is provisioned to the Terminal via Terminal Provisioning function.

For the above mentioned second case the terminal should support OMA BCAST Management Object parameter '/<X>/SGServerAddress/' as specified in [82]. Furthermore the entry point information may be fixed in the Terminal or provided out-of-band by the means such as WAP, SMS, MMS, Web page, user input, etc.

5.3.4 Scenario of using Single Service Guide to Provide Service Description for Multiple Service Providers

One important scenario of the ESG is that different operators want their customers get access only to their own ESGs. The basic idea is that the terminal gets the corresponding "Provider ID" (which corresponds to the relevant operator) in the ESG bootstrapping stage. Then the different "Provider ID" directs the access of the terminal to the relevant ESG transmitted from different operators.

In OMA BCAST, the association between the service providers and the individual fragments is provided using the grouping method of SGDD.

5.4 OMA BCAST BMCO Profile

The Broadcast Mobile Convergence Forum (BMCO Forum) is an international organization of companies targeting to shape an open market environment for mobile broadcast services [138].

As the OMA BCAST 1.0 standard is still in the finalizing stage, the BMCO members tested a pre-version of the OMA BCAST 1.0 standard with some elements disregarded for simplicity and a quicker market deployment.

The current OMA BMCO version ESG was according to OMA-BCAST ESG 1.0 of 25.05.2007 tested and it is a subset of OMA BCAST Standard and is forward compatible.

Table 5.1 illustrates the omitted elements and attributes of the OMA BCAST service guide 1.0 regarding to PreviewData. From Table 5.1, it can be seen that the current OMA BMCO profile (25.05.2007) does not have preview features for Synchronized Multimedia Integration Language (SMIL), video, audio, picture and text. The complete omitted elements and attributes in the BMCO profile can be obtained from [138].

Table 5.1. Overview of omitted elements and attributes of the service guide: PreviewData

Omitted elements and attributes	Functions
SMIL	This provides SMIL media for preview.
Video	This provides Video media for preview.
Audio	This provides Audio media for preview.
Under Picture the PictureData and Alternative text	This provides the actual Picture data and Alternative as part of the ESG. Note: under the profile the Picture is referenced as an external file and not carried inline in the ESG. The profile assumes a certain support for media codecs, hence alternative text is not necessary.
Text	This provides plain text for preview.

5.5 ESG Sharing

ESG Sharing refers to the concept that the common parts of ESGs from different ESG providers (in most cases mobile operators) can be shared and transmitted by the ESG Bootstrap providers (in most cases broadcasters). Fig. 5.3 illustrates the scenario of ESG sharing.

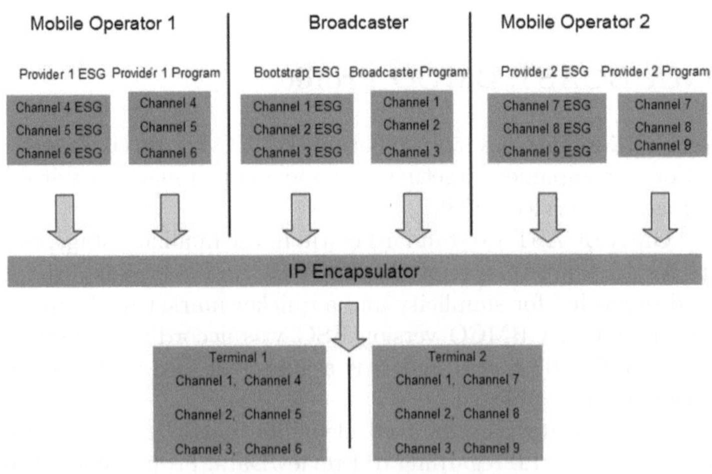

Fig. 5.3. ESG Sharing Scenario

As shown in Fig. 5.3, mobile operator 1 and mobile operator 2 utilize the DVB-H network infrastructure of the broadcaster. The broadcaster provide the bootstrap ESG. In Fig. 5.3 there are altogether 9 program channels being broadcasted to the terminals. Terminal 1 is the customer of mobile operator 1, so the terminal 1 is only allowed to receive the programs from mobile operator 1. For the same reason, terminal 2 is only allowed to receive the programs from mobile operator 2. Among the 9 programs, program 1, 2 and 3 are the common programs both both terminal 1 and terminal 1. While channel 4, 5, and 6 are only for terminal 1 and channel 7, 8, 9 are only for terminal 2. By using ESG sharing, the ESG for channel 1, 2 and 3 can be located at the broadcaster to save the overall bandwidth for both mobile operator 1 and mobile operator 2 instead of making the ESG for channel 1, 2 and 3 locate on both mobile operator 1 and mobile operator 2.

While both DVB IPDC ESG and OMA BCAST ESG have the ESG sharing concept, the ESG sharing concept in BCAST is also called "announcing service guides within a service guide".

There are also differences between the ESG sharing in DVB IPDC and in OMA BCAST. In OMA BCAST, a single BCAST ESG transport supports the marketing messages of several service operators; a separate ESG for each operator is needed in the IPDC ESG. In addition, the OMA BCAST ESG can be adapted to support both DVB-IPDC and BCAST terminals.

5.6 Comparison between DVB IPDC ESG and OMA BCAST ESG

The comparison between the DVB IPDC ESG version 1.0, version 2.0 and the OMA BCAST ESG version 1.0 are illustrated in Fig. 5.4.

As seen in Fig. 5.4, DVB IPDC ESG and OMA BCAST ESG defines different Conditional Access (CA) systems. Among the four different CA systems, Open Security Framework and OMA Smartcard Profile are smartcard based, while 18 Crypt and DRM Profile are device based security system. Although the three different kinds of ESGs all have the same components like ESG data model, encoding and encapsulation, DVB IPDC and OMA BCAST have different definitions for them. DVB IPDC ESG 2.0 also includes the features of unicast distribution of ESG just as OMA ESG does. However, this feature is not available in the DVB IPDC ESG 1.0.

One thing need to be known is the feature regarding to ESG Fragments compression. There are three ways to represent the ESG Fragments.

- Without compression;
- Compression using GZIP;
- Compression using Binary Format for Metadata (BiM)

Fig. 5.4 shows that OMA ESG 1.0 does not have the ESG Fragments compression mechanism of BiM, though it is claimed that BiM provides higher compression efficiency than GZIP regarding to ESG Fragments compression [92].

Features	DVB IPDC ESG 1.0	DVB IPDC ESG 2.0	OMA BCAST ESG 1.0
Conditional Access	Open Security Framework, 18 Crypt	Open Security Framework, 18 Crypt	DRM Profile, Smartcard Profile
Application Layer FEC	Yes	Yes	Not available
Notification	Not available	Yes	Yes
Data Model	ESG Data Model	ESG Data Model	SG Data Model
Encoding	ESG encoding	ESG encoding	SG encoding
Encapsulation	ESG encapsulation	ESG encapsulation	SG encapsulation
Unicast distribution	Not available	Yes	Yes
Interactivity	ESG based interactivity 1.0	ESG based interactivity 2.0	SG based interactivity
Fragment compression	GZIP and BiM	GZIP and BiM	GZIP

Fig. 5.4. Comparison between the ESGs of IPDC1.0, IPDC2.0, and OMA 1.0

5.7 Conclusions

The ESG is one of the most important parts within the DVB-H technology. On one hand it helps the users get the services and make interactions with the DVB-H operators. On the other hand, the DVB-H operators can utilize the ESG to control their customers, such as billing and locking their own ESG in the terminals.

The current ESG standard is development by two different groups, namely IPDC ESG and OMA BCAST ESG. Depending on who will be the DVB-H operator, different ESG standard maybe adopted. As OMA BCAST ESG defines the Smartcard profile and requires an uplink technology (for example 3G), it is supported by the mobile operators. While on the other hand, the IPDC ESG is mostly supported by the broadcasters whose customers include also unconnected devices (the terminal without telecommunication link). As a result, the IPDC ESG and the OMA BCAST ESG will most probably be parallel deployed in many countries.

Problems

5.1. How does pure OMA ESG work on one multiplex?

5.2. How is the Operator separation realized in DVB IPDC ESG?

5.3. How is the Operator separation realized in OMA BCAST ESG?

5.4. Does the ESG sharing concept exist in OMA BCAST ESG?

5.5. What are the interactivity features among different ESGs (IPDC1.0, IPDC2.0, BMCO OMA ESG, Standard OMA ESG?

5.6. How do the purchasing mechanisms work in different ESGs?

5.7. How do the mobiles handle different ESGs?

5.8. How are the SDP files being transmitted in different ESGs?

5.9. What are the EPG and multicast IP addresses in FLUTE for OMA ESG?

5.10. What is the port number in IPDC ESG bootstrap process?

5.11. What is the purchase interface in OMA ESG, is it HTTP or SMS or other interface?

5.12. What are the different ways to compress the ESG Fragments?

6

Handover Algorithm for a Dedicated DVB-H Network

The handover for the dedicated DVB-H network in this book is defined as the handover which is considered in the DVB-H network only. For the converged DVB-H and telecommunications networks, the handover inside the DVB-H network part can also be considered the handover of the dedicated DVB-H network. Handover in unidirectional broadcasting networks like DVB-H is a novel issue. The main difference between handover in DVB-H and that of 3G telecommunications networks is that passive handover in DVB-H is performed and can be performed by the terminals only while the handover in 3G telecommunications networks will require the operation by both the network and the terminals. Since making accurate handover decisions can reduce the battery power consumed, this chapter describes and investigates different strategies that can assist the handover decision-making process in DVB-H networks. The benefits and drawbacks of the different algorithms are presented. A hybrid handover decision-making algorithm is also described.

6.1 Introduction

Handover is the switching of a mobile signal from one channel or cell to another. When a user moves from one DVB-H cell to another, the DVB-H terminal has to be synchronized to another signal without service interruption. This chapter defines handover in DVB-H as a change of transport stream and/or frequency. Handover in dedicated DVB-H network refers to the handover in a DVB-H only network (without uplink connection).

DVB-H transmits data streams using a burst mode called time slicing instead of a continuous mode. Time slicing is the characteristic that makes seamless soft handover in DVB-H possible. The off burst time in time slicing transmission mode is illustrated in Fig. 6.1.

Depending on the transmission bit rate, the duration of the off time in the transmission stream can vary. The DVB-H receiver can utilize the off time to detect the signals and to initialize soft handover when it moves from one

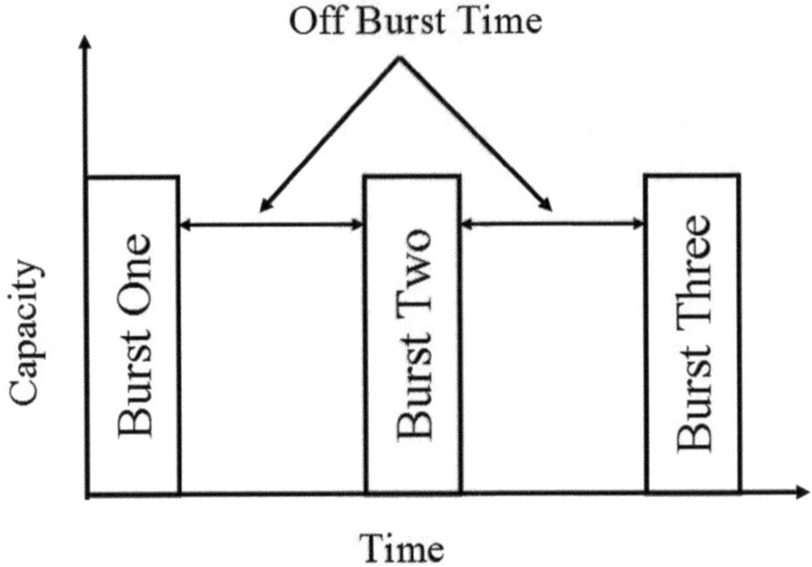

Fig. 6.1. Off Burst Time in Time Slicing Mode

DVB-H cell to another. This chapter focuses on the most important of the three handover stages - the handover decision-making process.

An instantaneous RSSI (Received Signal Strength Indication) value based handover scheme was proposed in [104]. This is the earliest publicly available handover algorithm for DVB-H. Since the RSSI value can vary due to multipath, interference or other environmental effects it may not give a true indication of communication performance or range and mistakenly measuring the RSSI value would result in unnecessarily consuming battery power because of more off burst time used in handover measurement. Therefore, the RSSI value may have the chance of being measured during many off burst times and the possibility is that the RSSI value would be measured at least every off burst time. Approaches to improve the RSSI handover algorithm are proposed in this chapter. In the analysis to the potential battery power consumption savings of the algorithms proposed in this chapter a worst case scenario for the RSSI value method is assumed that the RSSI value is measured every off burst interval.

A key idea in designing a soft handover algorithm for DVB-H is to predict the handover moment to reduce the handover measuring frequency in order to save battery power. The advantages and drawbacks of the proposed algorithms are presented and the comparisons between different algorithms are given. Based on this analysis, a hybrid handover decision-making algorithm is developed and future research directions for handover in DVB-H are suggested. The rest of the chapter is organized as follows:

Section 6.2 presents different handover decision-making strategies and their benefits and drawbacks; some of the algorithms are evaluated using numerical simulation. The comparison between different algorithms will also be given in this section. In section 6.3 a hybrid handover decision-making algorithm is proposed. Its feasibility and benefits are shown. Section 6.4 concludes the chapter.

6.2 Handover Decision-making Algorithms

In this section, novel handover decision-making algorithms are proposed and investigated. Their benefits and drawbacks are presented.

6.2.1 Context Aware Handover Decision-making

Since the handover area is usually the border area between cells, the context aware handover algorithm tries to predict the time it will take the terminal to move from its current location to the border area. In this way, the terminal can know the moment it should make handover measurement thus reducing the time spent taking handover measurements. Two typical scenarios are considered, an urban scenario and a rural scenario. Suppose the radius of the DVB-H cell is R(km), the distance between terminal and the transmitter station is L(km), and the velocity of the terminal is V(km/sec). Then the time it takes for the terminal to move from its current location to the handover area is given by:

$$T = (R - L)/V \tag{6.1}$$

R can be known in advance and obtained in the network planning stage. L can be roughly calculated from the signal strength measured by the receiver at the current location and the signal loss properties of the transmitted power. V can be estimated according to the specific environment, urban or rural. This is illustrated by an example shown later in this section. Namely, the V can be thought below a certain speed limit depending on the environment e.g., urban or rural. Such parameter information about the environment can be broadcast in the cell periodically so that when the terminal enters a cell it gets this information. The terminal can then determine the signal strength after the time interval T instead of measuring RSSI constantly. If the time between successive bursts of interest is denoted by $\triangle t$, since the RSSI handover scheme takes measurements every $\triangle t$ seconds and usually $T >> \triangle t$, the number of measurements needed for handover will be far less in the context aware handover scheme.

The graph illustration of this algorithm is shown in Fig. 6.2 and the steps that need to be taken to implement the context aware handover decision-making algorithm are:

Step 1: The DVB-H receiver extracts its environment context information from the received service streams.

Step 2: According to the environment context information (indicating rural or urban) received, the DVB-H receiver estimates its velocity and the DVB-H cell radius.

Step 3: The DVB-H receiver estimates its distance from the transmission base station of the cell it is located in.

Step 4: Using equation (6.1), the DVB-H receiver calculates the time interval T to perform soft handover measurement.

As the environment context is divided into urban and rural scenarios only, the handover measurement interval T will not change frequently keeping the power required low. If the environment context is divided into more categories, the handover will be more accurate but more battery power will be consumed because of the increased computing complexity.

A numerical simulation was used to evaluate this algorithm as follows:

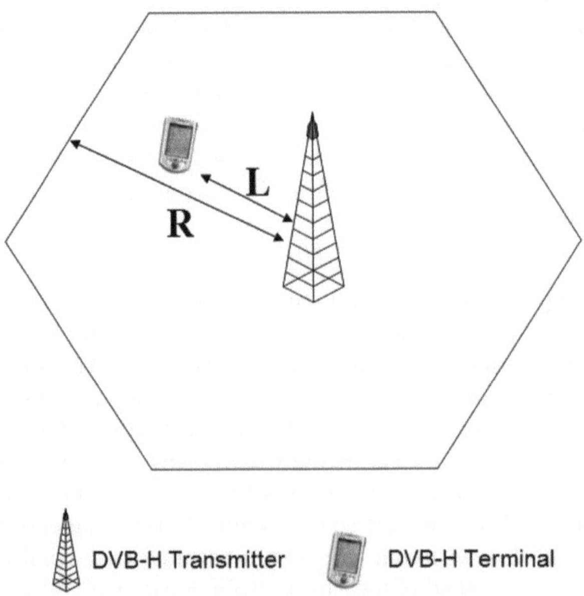

Fig. 6.2. Context Aware Handover

In the urban scenario, suppose the user is in a vehicle and the vehicle's velocity is usually below 48 km/hour [155]. Suppose the typical DVB-H cell radius in an urban area is 20 km, then the average value of $T = [20/(2\times48)]\times 3600 = 750$secs. In the rural scenario, the user's velocity is usually about 96 km/hour [22]. Suppose the typical DVB-H cell radius in a rural area is 40 km, L is half the radius to indicate an average distribution of users in the cell, then the average value of $T = [40/(2\times 96)]\times 3600 = 750$secs.

So typically the DVB-H receiver needs only measure the RSSI about every 750secs. This period is much longer than the off burst time that is at most a few seconds [104].

Consider a worst-case example. A user is driving along a road in rural area. The user's speed is 112km/hour. Suppose the user is using a DVB-H receiver and the radius of the DVB-H cell along the road is 40km. Suppose the user switches on the DVB-H receiver near the cell border area, for example 39.8km from the DVB-H transmitter base station. This DVB-H receiver's measuring interval is $[(40-39.8)/112]\times 3600 = 6.4$ sec.

The off burst time depends on the service being used and the burst bit rates. According to [104], the typical off burst time is about 3 seconds. The burst duration is usually much shorter than the off burst time. Suppose the burst duration is less than 1 second and the RSSI value is measured every off burst time, in this case, the RSSI handover scheme measuring interval $\triangle t$ will be less than 4 seconds.

Using the context aware handover decision-making algorithm, in the worst case example described above, about $[(6.4-4)/4]\times 100\% = 60\%$ of the battery power used by the RSSI value based scheme is saved; Because of the low computational complexity of equation (6.1), the context aware handover decision algorithm will consume very little power in the handover decision-making process.

The drawbacks of this algorithm are that the estimation of distance and velocity may not be very accurate. So this single algorithm cannot adapt to complex environments.

6.2.2 Location Aided Handover Decision-making

This algorithm uses mobile location information to aid the handover decisions in DVB-H especially in the motorway scenario.

The handover scenario considered is illustrated in Fig. 6.3.

Fig. 6.3 denotes a car running along a motorway from right to left. Suppose DVB-H transmitters are located along the motorway and that the DVB-H cells cover the entire motorway. It is assumed that the receiver will not receive stronger signals from a cell other than the cell in which it is located except in a cell border area. These assumptions make it easy to describe the model and do not affect the algorithm. Here it is assumed that A, B, C, and D in Fig. 6.3 are the points where the DVB-H receiver inside the car performs the handover.

The main idea of this location aided handover decision algorithm is that the handover will not be initiated until the car reaches a handover position,

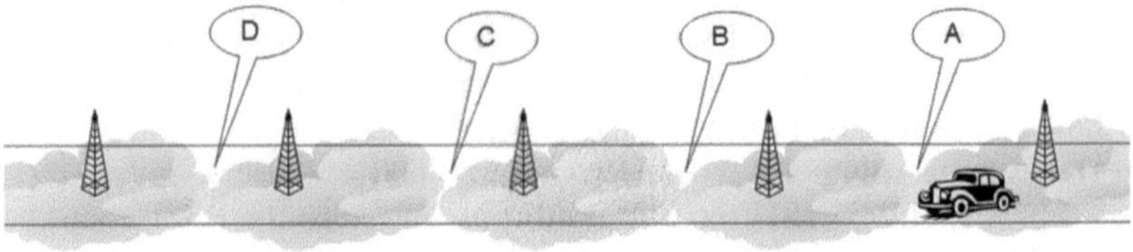

Fig. 6.3. Handover for DVB-H on Motorways

that is, A, B, C, or D in Fig. 6.3. In this case, the receiver will not use any of the off burst time for measurement until it reaches a handover position thus staying inactive most of the time and saving battery power.

The crucial question here is how the receiver can get to know whether it is in a handover position or not. Different techniques can be used to provide location information to the receivers. The most widely used location technique is Satellite Positioning Systems (GPS, GLONASS, Galileo) [48], [26].

The evaluation of this algorithm is as follows:

The vehicles on a motorway can be assumed to be an M/M/1 queuing model because the following assumptions are met:

1. Total number of vehicles driving on the motorway is very large.
2. A single vehicle uses a very small percentage of the motorway resources.
3. The decision to join the motorway is independently made by each vehicle.

The above observations mean that assuming a Poisson arrival process will be a good approximation of the vehicles passing the handover positions on the motorway. Although the distance between successive handover locations can be assumed to be constant because of the uniform motorway features, the individual vehicle speed is fluctuating because of the varying traffic. So the time interval between successive handover positions is variable. In this handover decision algorithm, the receiver making measurements can be modelled using a simple M/M/1 queuing model as shown in Fig. 6.4. In this model, the vehicle can be taken as the server; the handover positions can be taken as unlimited customers arriving at the server randomly.

The states of the queuing system are assigned to discrete handover locations along the road. The number of cells along the motorway route determines the number of states in the model.

The transmission probability of the model λ can be calculated using the average vehicle speed v and the distance between successive handover positions $l : \lambda = v/l$. The average vehicle speed v can be obtained according to the speed limit of the road. The distance between successive handover positions l can be obtained according to the cell diameter. It is assumed without loss of generality that the vehicle travels on a fixed route and the bi-directional traffic is represented by two models one for each direction. In Fig. 6.4, the state

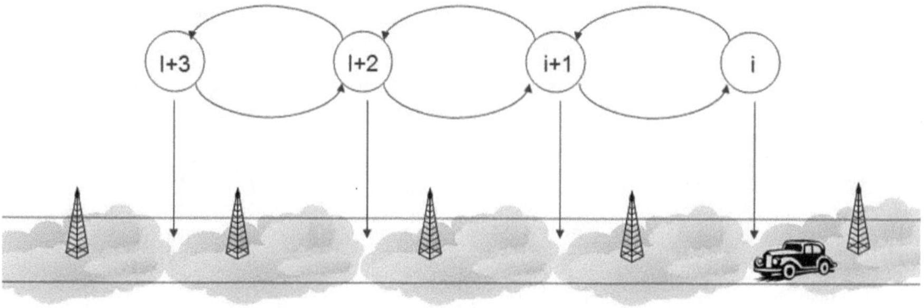

Fig. 6.4. Handover Algorithm Based on M/M/1 Model

transition of our model is only from right to left. In this model, the probability of arrival at a handover location is the probability of the vehicle reaching a new cell.

Suppose that the average cell diameter is $l = 10$ Kilometers and the speed limit on the motorway is $v = 90$ kilometers/hour. Then $\lambda = 90/(10 \times 3600) = 0.0025$. The average inter arrival time will be $T = 1/\lambda = l/v = 400$seconds. Because the time the user is in the handover position, that is the service time, is very short, suppose it is 10 seconds, then this algorithm will save up to $400/(400+10) = 97\%$ of the battery power consumed by measuring the RSSI value every off burst interval. It is easy to see that the larger the cell size and the slower the vehicle speed, the more of the battery power used by the RSSI algorithm will be saved.

The drawback of this algorithm is that incorporating Satellite Positioning Systems components into the handset will make the handset more expensive and usually Satellite Positioning Systems do not provide location assistance indoors. Thus, this algorithm is only suitable for the motorway scenario or vehicle based DVB-H receivers.

6.2.3 UMTS Aided Handover Decision-making

Handover decision-making in DVB-H can be assisted by the terminal's UMTS link if a converged UMTS and DVB-H network serves the terminal. This kind of converged network structure is described in detail in [27]. The converged network structure is taken from [27] and shown in Fig. 6.5. In this scenario, there are always UMTS base stations (Node B) located in the DVB-H cell border area under the assumption that one DVB-H cell covers several UMTS cells. This assumption is based on [159] which claims that the use of small size cell is not an advantage in broadcast network planning.

When the UMTS/DVB-H terminal moves to the DVB-H cell border area which is the handover position area, it will receive the information from the UMTS base stations so that it knows that the DVB-H handover measurement should be performed.

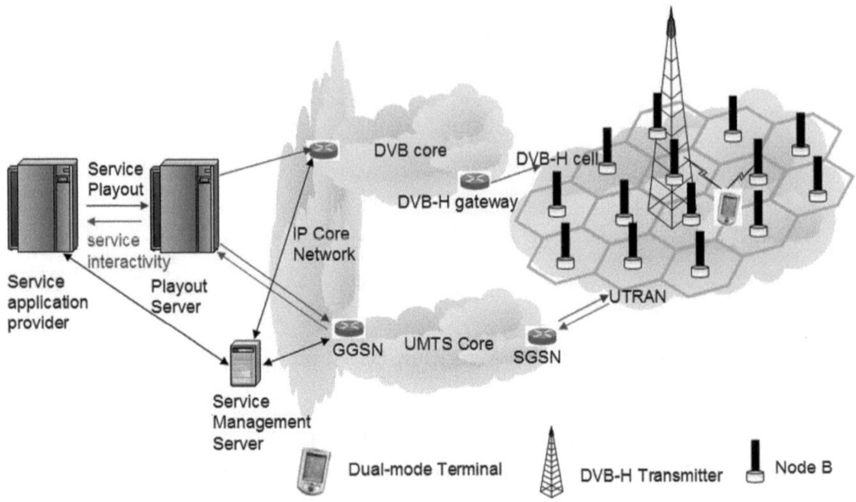

Fig. 6.5. Converged DVB-H and UMTS Network Structure

The performance of this handover decision-making algorithm depends on the performance of the UMTS connection of the terminal. The terminal moving into the DVB-H handover area also makes handover in the UMTS network from one cell to another compulsory. If the UMTS cell that is in the DVB-H handover area is too crowded the UMTS handover request may be blocked when the terminal moves into the handover area. The handover failure probability is the main performance criteria of this algorithm. To avoid handover failure, cell broadcasting [156] can be used to provide the handover measurement initialization information to the terminals that move into the handover area. The battery power consumption reduction compared with the RSSI algorithm depends on the UMTS cell size in the handover area compared with the DVB-H cell size. It is obvious that the smaller the UMTS cell size and the larger the DVB-H cell size, the more battery power consumption can be saved.

The drawback of this algorithm is that the DVB-H handover accuracy and reliability depends solely on the UMTS base station in the handover area. So the algorithm's complexity is increased.

6.2.4 Repeater Aided Handover Decision-making

Repeaters have been important network components for both analogue and digital TV broadcasting [116]. In the planning and optimization of DVB-H networks, a repeater could be used to extend the transmitter coverage area or to cover a shadow area, such as a tunnel, valley, indoor area, etc. There are usually repeaters in a cell border area. In this handover decision-making algorithm intelligent repeaters are used to provide location information to the receivers. When such repeater specific information is delivered to the receiver within the transport streams, the receiver will perform the handover measurement at the right moment. Details of this algorithm are presented in chapter 8 of this book where it is shown that using this algorithm up to 63.22% of the average front-end battery power consumption of measuring RSSI values could be saved.

The main drawback of this algorithm is that new intelligent repeaters must replace the old repeaters in the cell border areas. The expenses of installing such new repeaters must be considered by the network operator.

6.2.5 Other Handover Decision-making Algorithms

Since the key idea of all the different handover decision-making algorithms for DVB-H proposed in this chapter is to predict the handover measurement moment, the more accurately the handover measurement moment is predicted the better the handover decision-making algorithm is. There are some other handover decision-making algorithms potentially available. Pattern recognition is an example of a technique that could be used to assist in making accurate and timely handover decisions. A Hidden Markov Model (HMM) based algorithm is proposed here as an example of the possible use of pattern recognition techniques in handover decision-making in DVB-H.

Similar Hidden Markov Models have been proposed for cellular GSM networks [125]. The proposed algorithm utilizes Hidden Markov Models trained with previously collected data to model the strength of the received signals for different DVB-H cells. The strength of the received signals is measured from the received service signals without occupying the off burst time. Then the terminal uses the received signal strengths to decode the Hidden Markov Model of the cell it is located in. When the terminal is near the cell to which it is moving into, it perceives the change in the Hidden Markov Model. Then it makes the decision to perform the handover measurement. The basic idea of this handover decision-making algorithm is illustrated in Fig. 6.6.

This algorithm requires no modifications to existing standards or DVB-H handsets. Therefore, it may lead to cost effective, reliable solutions. Since this algorithm is based on the previously collected data that are the received signal strength measurements, the prediction precision is the most important factor for its success. The drawback of this algorithm is that when the terminal is idle most of the time, it cannot get enough measurement data for accurate model prediction.

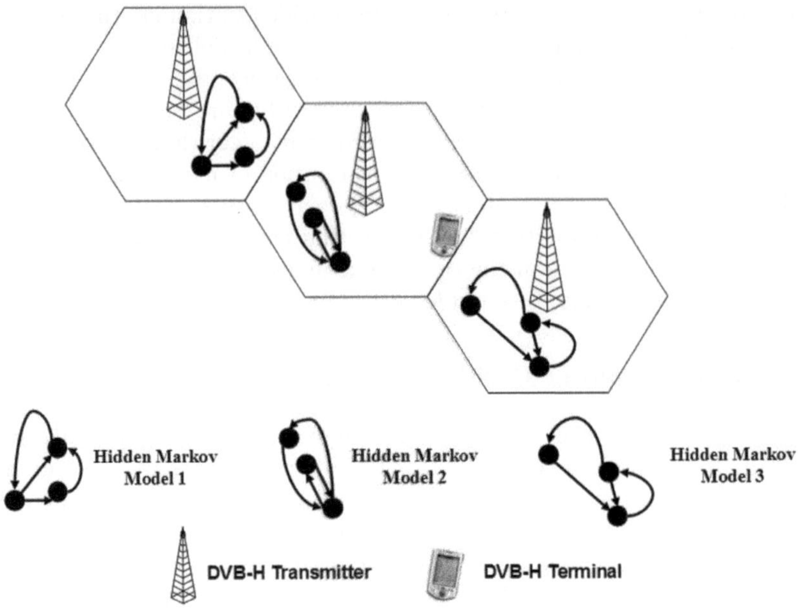

Fig. 6.6. Hidden Markov Model Based Decision-making Algorithm

6.3 Comparison of Different Handover Decision-making Algorithms

The power consumption of the proposed algorithms is compared with RSSI algorithm. The comparison is shown in Table 6.1.

6.4 Hybrid Handover Decision-making Algorithm

Implementing one handover decision-making algorithm has the limitation that the algorithm may work well in one specific environment but not all environments. The future mobile service is an anytime anywhere service. The terminal will be used in all kinds of environments. In this case, implementing one algorithm cannot cope with all the situations. A hybrid handover decision-making algorithm is thus proposed as described below.

The basic idea of this hybrid algorithm is that a central management module manages the different algorithm modules installed in the terminal as illustrated in Fig. 6.7.

The hybrid algorithm is:

Step 1: The DVB-H receiver extracts environment context information from the received service streams.

Table 6.1. Comparisons Between Different Algorithms

Algorithms	**Saved power consumption compared with RSSI algorithm**	**Advantages**	**Disadvantages**
Context aware	60% in the worst case	Simple and efficient	Less robust to environment
Location aided	Up to 97%	Simple and efficient	Costly and only in motorway scenario
UMTS aided	Not determined	Simple and efficient	Complex and needs UMTS network
Repeater aided	Up to 63.22%	Simple and efficient	Costly
HMM based	Not determined	Simple, less costly, efficient	Needs enough measurements data

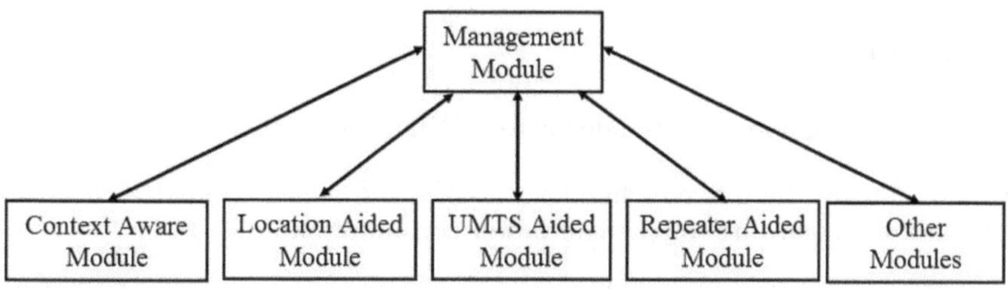

Fig. 6.7. Hybrid Algorithm Modules

Step 2: According to the environment context information, the DVB-H receiver chooses a handover decision algorithm as follows:

1. The management module chooses an algorithm at random.
2. The performance of the chosen algorithm is evaluated and a score or grade is assigned by the management module.
3. When the receiver is switched on the next time one of the algorithms that have not yet been used is selected at random and step 2 applied.
4. After many algorithm evaluations, the best performing algorithm for each environment is identified. These algorithms are chosen as the default algorithms for the respective environment.

In this hybrid algorithm, the environment information is divided into different categories, urban residential area, rural residential area, pedestrian area, motorway area, etc. These detailed categories of different environment need to be defined and evaluated in real field trials. The evaluation of this hybrid algorithm is the subject of future research.

6.5 Conclusions

The critical phase in the DVB-H handover procedure is the handover decision-making process. A good handover decision-making algorithm can greatly save consumed handset front-end battery power. Novel DVB-H handover algorithms have been described and investigated in this chapter. However, the different algorithms have different limitations. A hybrid handover decision-making algorithm has been proposed to utilize the advantages of the different algorithms while avoiding their limitations. Although this hybrid algorithm has not been evaluated by test, its potential feasibility in real environment has been made clear. The validation of the different handover algorithms should be done in the future field trials.

Problems

6.1. What is handover in the dedicated DVB-H netowrk?

6.2. What are the different handover stages and which one is the most important one regarding to terminal power consumption?

6.3. What is the key idea to design a soft handover algorithm for DVB-H?

6.4. Why do we need hybrid handover decision-making algorithm?

7

Post Processing of SNR Based Handover

In this chapter a novel soft handover mechanism for DVB-H is described which is based on measuring the Cumulative Distribution Function (CDF)of the signal to noise ratio (SNR) by the DVB-H terminal receiver front-end. Details of the algorithm are given and simulation was done to prove the benefits of such a soft handover scheme.

7.1 Introduction

In this chapter, a novel soft handover algorithm for DVB-H is proposed and some simulations using OPNET [57] and Matlab are presented. Although a seamless handover is difficult to be realized in reality, the post processing of SNR based Handover algorithm can be assumed as seamless handover where the handover only happens within the off burst time during the DVB-H time slicing cycle.

7.2 Description of the Algorithm

Dohler [114] presented an interesting idea that described a simple power drop model based on a distance dependent time-gradient for handover in cellular telecommunications networks. However, this model is based on the base station side of a bi-directional cellular telecommunications network. In this chapter, seamless soft handover is proposed based on the post processing of the SNR (Signal to Noise Ratio) instead of the RSSI; thus avoiding frequent handovers as will be shown in this Chapter. The SNR is calculated from the RSSI and the noise characteristics and thus provides a more accurate estimate of the received effective signal.

Suppose that all the service information and network information are already stored in the memory of the receiver, this chapter focuses on the handover measurement and decision. The proposed algorithm is illustrated in Fig. 7.1 below.

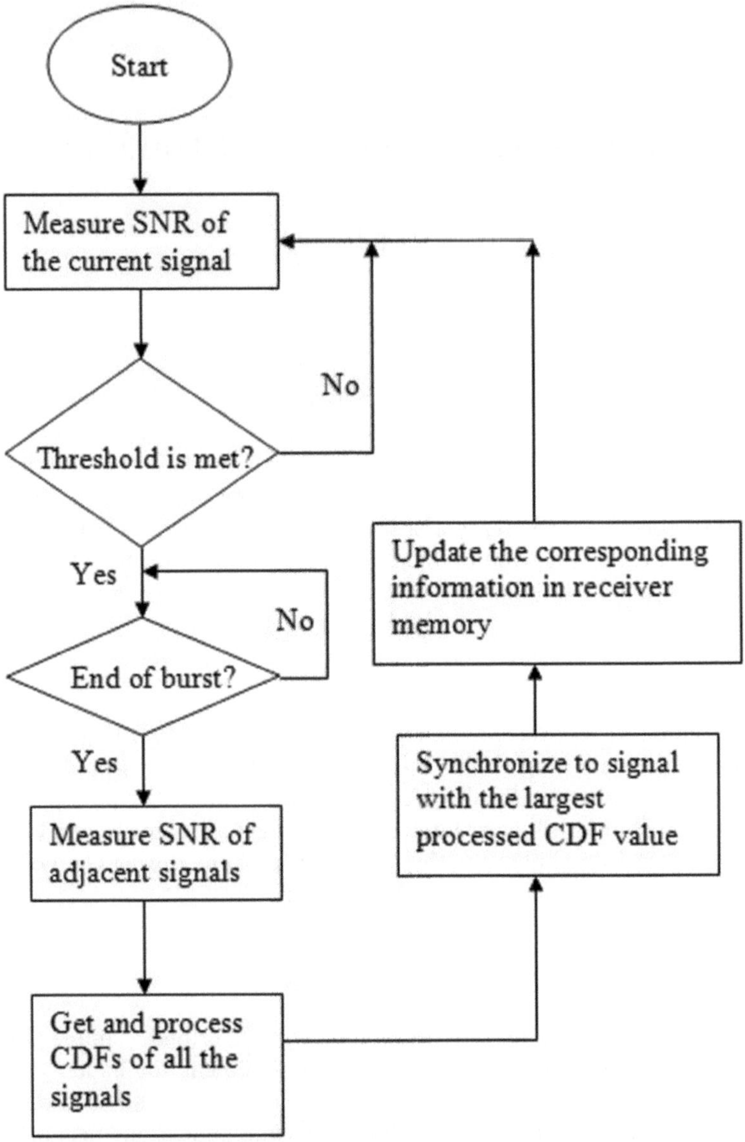

Fig. 7.1. Seamless Soft handover algorithm

In Fig. 7.2, when the receiver gets all the SNR values of the adjacent signals, it will calculate the CDFs (Cumulative Distribution Functions) of all the SNR values. A Cumulative Distribution Function describes a statistical distribution. It gives at each possible outcome the probability of receiving that outcome or a lower valued one. The computation of CDF is shown in equation (7.1).

$$F(x) = P(X \leq x) = \int_{-\infty}^{x} f(x)dt \tag{7.1}$$

where $f(x)$ is the probability density function for random variable X, and $F(x)$ is the probability of observing any outcome less than or equal to x.

Because the CDF gives a probability value, its value depends not only on the current SNR but also on the SNR history of the signal. This not only eliminates the frequent handover phenomenon seen in instantaneous RSSI value based handover but also avoids the "fake signals" [104] caused by frequency confusion. The "fake signals" can be eliminated because the "fake signals" are only caused by short period signals and evaluating the signals' history reduces the chances of making decisions on short period signals . The simulation presented below shows that the CDF based handover decision is seamless and more reliable than RSSI based handover.

7.3 Simulation and Analysis

Simulation is done to illustrate and test the soft handover algorithm for DVB-H proposed above in this chapter. The simulation was done using OPNET and the results were analyzed using MATLAB.

A simple DVB-H model was constructed in OPNET which is shown in Fig. 7.2.

The area of southwest of Britain was chosen as the terrain model background because it contains various geographical features: plains, open spaces, hilly and mountainous rural areas, rivers, seas, etc. These complex terrain features make the simulation more realistic. The Longley-Rice propagation model was used to compute the signal path loss [151]. The Longley-Rice model is also known as the Irregular Terrain Model (ITM). It is intended to be used for radio frequencies from 20 to 20,000 MHz and for distance less than 2000 Km. It is very suitable to be used in OPNET for the terrain models.

In the scenario shown in Fig. 7.2, for simplicity two DVB-H transmitters (dvb_station_0 and dvb_station_1) are placed in two different mountainous areas. The DVB-H receiver moves along the black curve track (shown in Fig. 7.2) from the southwest tip of the area towards northeast. As the receiver moves, it measures the SNR from the two different transmitters. Because the receiver's movement is irregular, which matches very well with the reality, the measured SNR values for the signals from the two transmitters will fluctuate.

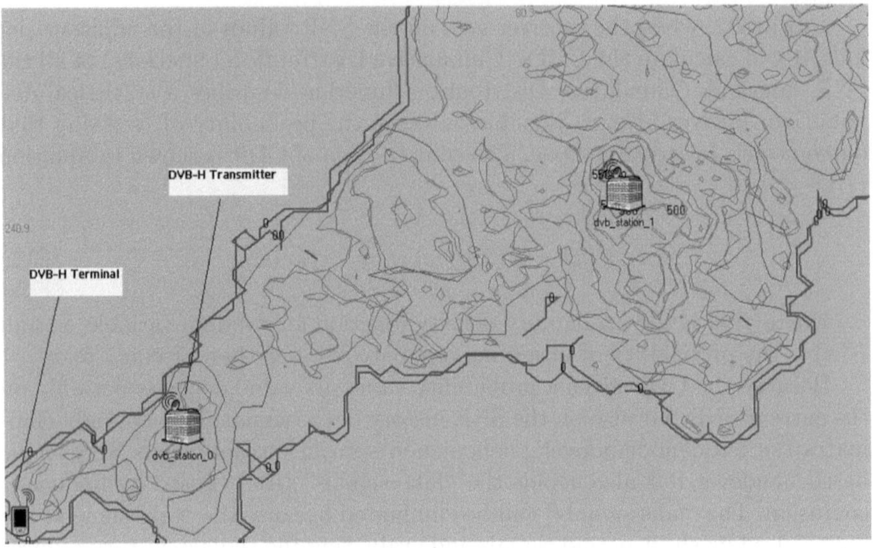

Fig. 7.2. Soft Handover Scenario

After 15 hours simulation time, the SNR statistics from both transmitters were obtained and are shown in Fig. 7.3.

In Fig. 7.3, the different lines are standing for different SNR measurements of the signals from dvb_station_0 and from dvb_station_1. It can be seen that the SNR measurements of the signals from dvb_station_1 are above the SNR measurements of the signals from dvb_station_0 around point A in Fig. 7.3. If the handover decision is based on the instantaneous value of SNR, then the handover to dvb_station_1 will happen at around point A but it will soon handover back to dvb_station_0 according to the SNR plots. This will cause unnecessary frequent handovers that is known as the *Ping Pong* effect. To avoid such unnecessary frequent handovers, post-processed CDFs of the two SNR curves are used to make more accurate handover decisions.

The data of the simulation results done in OPNET were imported to MATLAB to compute the CDFs of the two SNR curves using a uniform distribution. As the discrete points of a CDF curve are not suitable for making a handover decision, the Savitzky-Golay method [49] was used to smooth the CDF curves because of its better accuracy than the other smoothing methods. The results of this exercise are shown in Fig. 7.4.

In Fig. 7.4, the lines for the smoothed CDF for dvb_station_0 and the smoothed CDF for dvb_station_1 are indicated in the figure. It can be seen that the moment in time corresponding to point B (the crossing point between the smoothed CDF for dvb_station_0 and the smoothed CDF for dvb_station_1) should be the time to carry out handover from dvb_station_0 to dvb_station_1. Because such handover happens during the off burst time

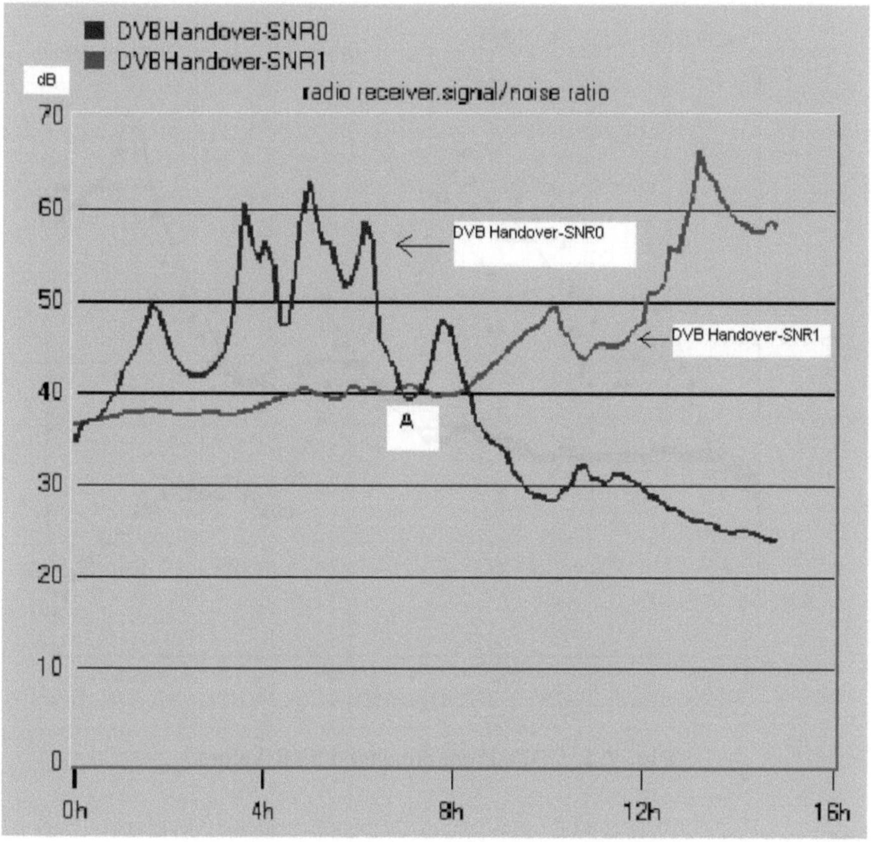

Fig. 7.3. SNR From Transmission Stations

so it can be regarded as seamless, and because it is not dependent on the instantaneous SNR values, the "fake signals" scenario will not happen.

7.4 Conclusion

In this chapter, a seamless soft handover scheme for DVB-H is described, which is also a handover algorithm for the dedicated DVB-H networks. This algorithm is based on the post processing of the CDFs obtained from the SNR values at the receiver front-end. The proposed algorithm can effectively eliminate frequent handover and avoid the "fake signals" caused by RSSI based handover. A simulation was presented to illustrate the benefits of the proposed algorithm. The same as other DVB-H handover algorithms, this algorithm has to be validated in the real field test.

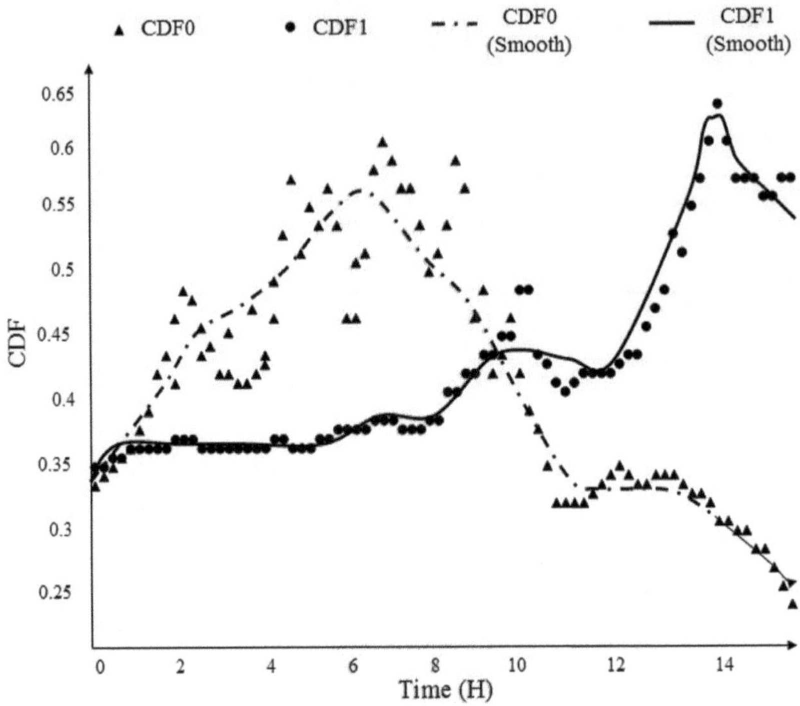

Fig. 7.4. CDFs From Simulated SNR Values

Problems

7.1. What is the advantage of choosing SNR instead of RSSI as handover measurement criteria?

7.2. What is the advantage of calculating the CDFs of the measured SNR values in the handover algorithm?

8

Repeater Aided Soft Handover

8.1 Introduction

This chapter proposes and analyzes the Repeater Aided Soft Handover (RA_handover) algorithm for a DVB-H receiver with Multiple-Input Multiple-Output (MIMO) antennas and presents the benefits of implementing the RA_handover compared with the handover process without repeaters. For network planning and optimization purposes simulation models are developed to analyze the RA_handover approach. It is shown that RA_handover could greatly improve the quality of service and consume much less front-end battery power than the handover method without repeaters. In addition, the costs of implementing the algorithm are briefly estimated. In conclusion, curves are given that have shown the relationship between the quality of service and the consumed battery power, which gives further justification for the repeater aided handover to be included in the DVB-H soft handover standard.

Repeaters provide an efficient solution to increase the coverage of the broadcasting networks [72]. In broadcasting networks, the network operators usually firstly put high power transmitters at strategic points to quickly ensure an attractive coverage and then, at a second step, increase their coverage by placing low power repeaters in the dead spots or shadow areas, such as tunnels, valleys, or indoor areas. A repeater is simply a device that receives an analogue or digital signal and regenerates the signal along the next leg of the medium. In DVB-H networks, there are two different kinds of repeaters. They are passive repeaters which are also called gap-fillers, and active repeaters that are also called regenerative repeaters. A passive repeater receives and retransmits a DVB-H signal without changing the signalling information bits. The signal is only boosted. An active repeater can demodulate the incoming signal, perform error recovery and then remodulate the bit stream. The output of the error recovery can even be connected to a local remultiplexer to enable insertion of local programs. It means that the entire signal is regenerated. The building blocks of the passive and active repeater configurations are shown in Fig. 8.1. The repeaters used in RA_handover approach are active repeaters.

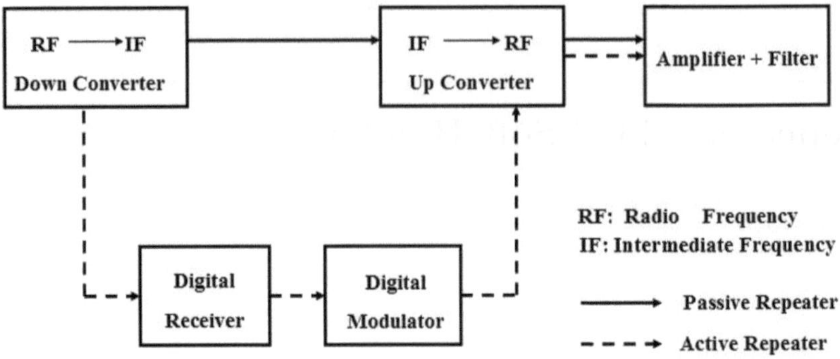

Fig. 8.1. Building Blocks of Passive and Active Repeater Configurations

This chapter describes a novel approach called RA_handover to decide when soft handover should occur by incorporating intelligent digital repeaters into the DVB-H networks. The DVB-H terminal used in this chapter is DVB-H capability only receiver (the so called "unconnected devices") with MIMO antennas [28] so that the receiver can receive and process the signals from different transmitters and repeaters at the same time. In this handover approach an intelligent active repeater structure is proposed where each repeater can add repeater identification bits to the received DVB-H signal and retransmit it to the repeater covered area to provide location information to mobile receivers. Such an algorithm will greatly improve the quality of service of the received signals and reduce the receiver battery power consumption without considerably increasing the overall cost.

In this way, a receiver does not need to measure the handover parameters before it reaches the handover location reducing the *Ping Pong* effect and consequently battery power consumption. On the other hand, "fake signals" will be completely eliminated because all the repeaters provide their unique identification information to the receivers.

The chapter is organized as follows:

Section 8.2 describes the signalling bits proposed in the RA_Handover scheme. Section 8.3 gives a detailed description of the proposed approach to decide when soft handover should occur, namely, repeater aided soft handover or RA_handover. In Section 8.4, a simulation model is built and the performance analysis is done for the RA_handover approach. Section 8.5 concludes the chapter.

8.2 DVB-H Signalling For RA_Handover

To implement handover in DVB-H, the receiver needs to receive signalling information from the network. There are two kinds of signalling information the DVB-H receiver can use. One is TPS (Transmission Parameter Signalling) signalling bits in the physical layer [70]. The other is Service Information (SI) description data that forms a part of the DVB-H transport streams [72]. Yang et al. [25] proposed some new signalling information for TPS and SI in DVB-H soft handover, which are described briefly as follows:

TPS is defined over 68 consecutive OFDM symbols referred to as one OFDM frame. Each OFDM symbol conveys one TPS bit so each TPS block contains 68 bits [70]. The TPS bits needed for handover are derived from [70] and listed in Table 8.1.

The Synchronization Word bits in TABLE 8.1 aid the receiver in synchronizing with the target transport stream and/or frequency. The Cell Identifier in TABLE I conveys unique cell identification information to the receiver. Bits numbered $S_{48} - S_{53}$ in TABLE 8.1 were originally defined as *Reserved For Future Use.* All the other TPS bits are already used for certain functions in the DVB standard [70]. Some of these Reserved For Future Use bits could be used to realize the proposed RA_handover approach.

The SI data provide information on the DVB-H services carried by the different transport streams. Handover related information in SI is contained in the NIT (Network Information Table), which is derived from [70] and defined in TABLE 8.1.

Table 8.1. TPS Signalling Information for Handover

Bit number	Purpose/Content
$S_1 - S_{16}$	Synchronization word
$S_{40} - S_{47}$	Cell identifier
$S_{48} - S_{49}$	DVB-H signalling
$S_{50} - S_{51}$	Handover types
$S_{51} - S_{53}$	Reserved for future use

If the cell id information is announced in the TPS bits, the NIT (Network Information Table) in SI data will contain both a cell_frequency_link_descriptor and a cell_list_descriptor announcing all cells and subcells within the DVB-H network.

Using the TPS and SI information, the receiver can initialize and decide when handover should take place

8.3 RA_handover Algorithm

Before going into the details of the proposed RA_handover algorithm, the novel active repeater structure it requires is shown in Fig. 8.2.

In the repeater structure shown in Fig. 8.2, the Pseudo Random Binary Sequence (PRBS) Generator is an integral part of the digital repeaters. The

TPS adapter adds the unique repeater specific information to the TPS bits in the transport stream.

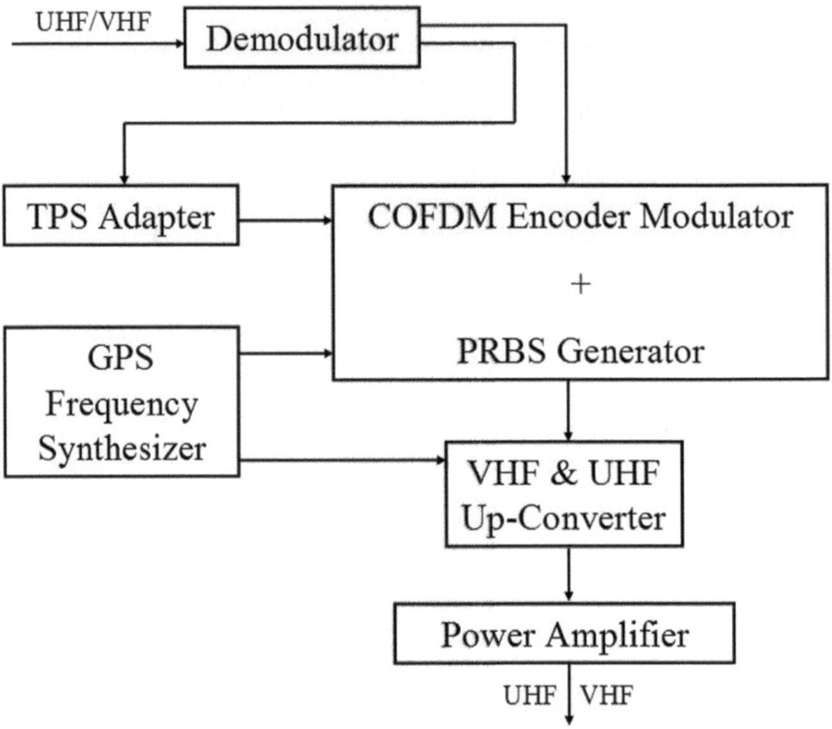

Fig. 8.2. Active Repeater Structure in RA_handover

In the RA_handover approach, the active repeaters are located in the cell border area. Each repeater-covered area is defined as one subcell. Unlike a passive repeater that simply amplifies and relays an incoming signal, an intelligent active repeater can demodulate the incoming transport stream, add handover scheme information and subcell id information to the TPS bits, and add subcell id information to the SI bits in the transport stream.

In the RA_handover approach, the DVB-H receivers have MIMO (Multiple Input Multiple Output) antennas that can provide better receiving and decoding capability from the different transmitters and repeaters at the same time than receivers without MIMO antennas.

For the proposed RA_handover approach, intelligent active repeaters are put uniformly around the cell borders in a cellular DVB-H network. Each repeater-covered area is called one subcell. When a mobile device moves into such a subcell it receives the unique repeater identification information from the repeater transmitted signals. Since the repeaters have unique identification information being transmitted in the transport stream they radiate, the mobile device will know in which specific subcell it is located. When the device is in a repeater covered subcell, it will begin to measure the signal strength using the off burst time. Otherwise the receiver is in sleep mode during the off burst time. In this way, the measurement frequency in the off burst time is greatly reduced, thus saving battery power and improving quality of service. In addition, it also reduces the *Ping Pong* effect.

The cellular network structure for RA_handover is shown in Fig. 8.3.

Fig. 8.3 shows a seven-cell DVB-H network topology. Each cell contains six repeaters, i.e. six subcells, allocated uniformly around the border of the cell. R_{12} is one subcell in cell 1 at the border between cell 1 and cell 2. R_{21} is one subcell in cell 2 at the border between cell 2 and cell 1. Correspondingly R_{ij} and R_{ji} are the subcells at the border between cell i and cell $j(i,j = 1,2,3,4,5,6,7)$ respectively. Suppose the repeaters are using directional antennas and each repeater can cover and only covers the subcell area where it is located. When the mobile receiver moves into any subcell area covered by a repeater, it will get the corresponding repeater information from the signalling bits it receives within the on burst time. At this location the receiver will begin to carry out handover measurements in the off burst time. This means that the receiver will not measure the signal strength using the off burst time until it reaches a subcell area covered by a repeater. In this way the receiver does not need to measure the signal strength level constantly saving battery power. With the installation of the repeaters in the cell border area, the quality of service will also be increased compared with that of no repeaters installed in the border.

With MIMO (Multiple Input Multiple Output) antennas on the DVB-H receiver, the receiver can receive signals from different directions and combine them into a better quality signal, thus improving the quality of service. Take cell 1 for example, as shown in Fig. 8.3, it is easy to see that another advantage of the RA_handover algorithm is that the receiver will always feel it is at the centre of the cell no matter wherever the receiver moves within the cell. In this way, the RA_handover algorithm can not only improve the quality of the service that the receiver received but also keep the quality of the service coherent all over the cells.

With the addition of repeaters, the main transmitter power can be reduced, thus reducing the operational costs of the main transmitter. Although the cost of the addition of the repeaters will add additional cost to the network equipment, the overall cost of the system will not be increased when the increased quality of service and the saved cost in the terminal side because of saved power consumption are considered. The cost issue will be considered in the future work.

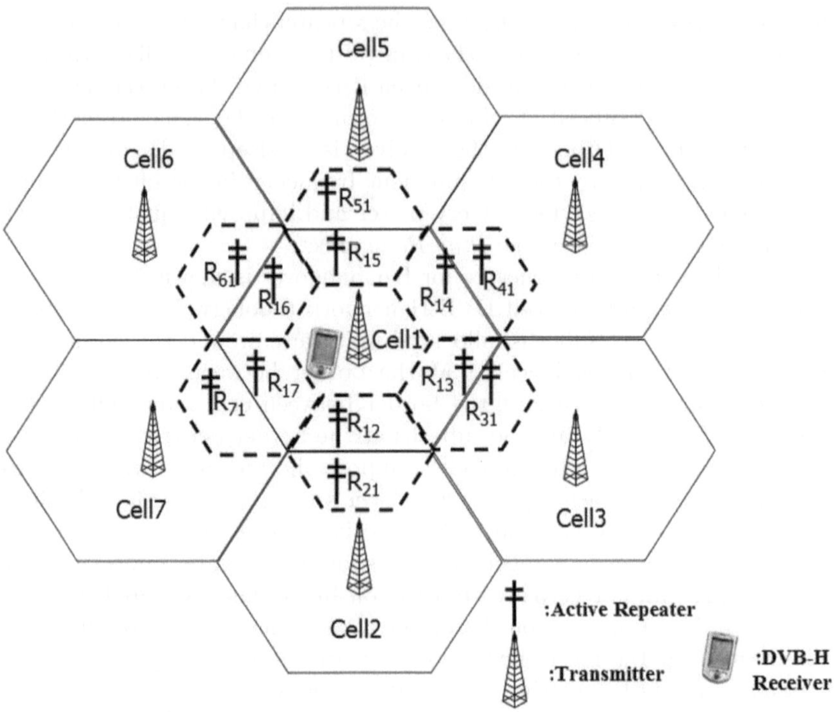

Fig. 8.3. RA_handover Algorithm Cellular Structure

8.4 Simulation Model and Analysis

The performance of the RA_handover algorithm will be analyzed in this section with respect to the front-end battery power consumption, the received quality of service and the cost of the overall network system. The approach of this section is to build a simulation model in MATLAB to identify the relationship between the received signal strength and the battery power consumption. The received signal strength is related to the repeater-covered area. The simulation scenario is that the repeater-covered area is changed as the received signal strength from the repeaters is changed. Cell radius, antenna height, transmitter power, transmission frequency and time percentage are their common parameters on which the received signal strength depends. Time percentage is a term widely used in propagation modeling, it accounts for variations in hourly median values of attenuation due to, for example, slow changes in atmospheric refraction or in the intensity of atmospheric turbulence. The value of time percentage gives the fraction of time during which the actual received field strength is expected to be equal to or higher than the hourly median field strength. This variable allows the time variability of changing atmospheric (and other) effects to be specified. As the received signal strength can be thought of as proportional to the received quality of service, the relationship between the quality of service and the battery power consumption can be obtained. The simulation parameter data were derived from the DVB-H standards [70, 73, 72] and International Telecommunications Union (ITU)standards [85].

Research has shown that human factors are essential in incorporating a successful service delivery system for wireless telecommunications [29]. Since cost is one of the human factors and it is a very important issue in business and standardization process [30], the cost issue is described analytically in the last part of this section.

First we calculate the percentage of battery power that can be saved using RA_handover algorithm compared with the algorithm in which every off burst time is used to make handover measurement that may happen without repeaters. From Fig. 8.3 it is easy to see that the receiver will only make handover measurement in the six subcell areas instead of the whole area of cell 1. The more handover measurements that are made the greater battery power consumption will be. The handover probability can be obtained from the area where the handover will happen and the whole service area [39]. By using the same methods, the saved power consumption can be calculated from the difference of the repeater covered area and the whole cell area. Suppose that the whole cell area and the repeater-covered area are ideal hexagonal shapes as shown in Fig. 8.4 and the DVB-H receiver is uniformly distributed in both time and location in the cell. Fig. 8.4 shows the maximum area the repeaters are able to cover. The following equation is obtained:

$$S = \frac{A_c - A_r}{A_c} = 25\% \qquad (8.1)$$

In equation (8.1), A_c is the area of the whole cell, A_r is the whole area covered by the six repeaters, S is the saved battery power compared with the handover algorithm utilizing every off burst time. Thus it can be seen that at least 25% of the battery power consumption on the handover decision-making stage can be saved.

It needs to be noted that in the network topology shown in Fig. 8.4 the receiver will receive the best quality of service because it always receives as if is near the centre of the cell. On the other hand Fig. 8.4 shows the maximum area that the repeaters cover. In this case the saved battery power consumption S is minimum. If the repeater covered area A_r is decreased, the saved battery power consumption S will be increased but the quality of service will be decreased too. Because the receiver will not receive as if it is near the centre of the cell again when A_r is decreased, the quality of the service will not be coherent all over the cell.

In order to determine the relationship between the battery power consumed by the handover decision-making algorithm and the received quality of service a model is built up for simulation. Without losing generality suppose

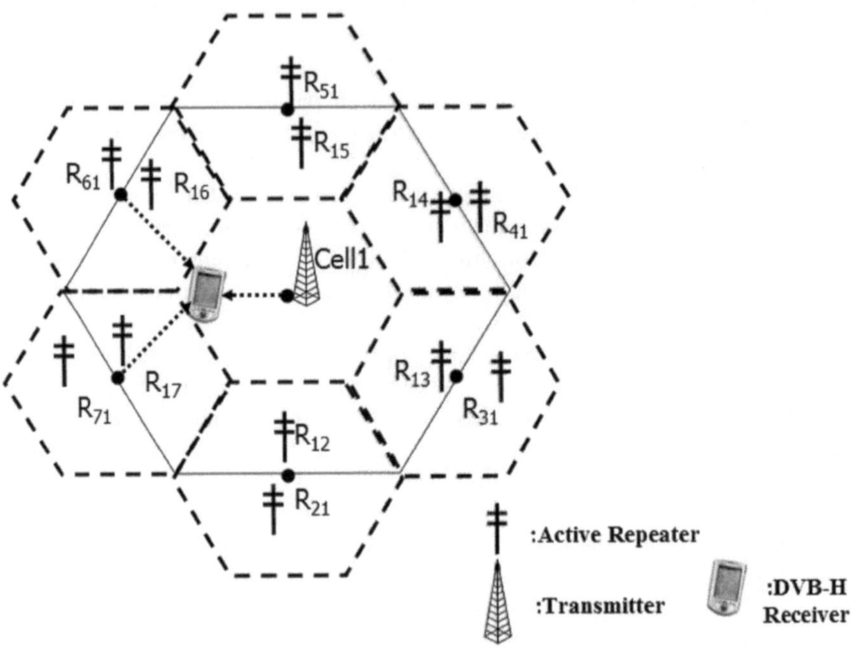

Fig. 8.4. The Receiver Always Feels It Is at the Centre of the Cell

the received quality of service is directly related to the received signal strength. Although the received quality of service and the received signal strength is nonlinear relationship, it is simpler and does not affect the purpose of the simulation to assume that there is a fixed linear relationship between the received quality of service Q and the received signal strength E_b:

$$Q = \alpha E_b \tag{8.2}$$

Where coefficient α is a constant parameter that links the Q and the E_b together.

Correspondingly for simpler simulation and without affecting the purpose of the simulation suppose that there is a fixed linear relationship between the battery power consumed by the handover decision-making algorithm C and the size of the repeater covered area A_r.

$$C = \beta A_r \tag{8.3}$$

Where coefficient β is also a constant parameter that connects the C and the A_r together.

The repeater covered area or the range of the repeaters depends on several things, such as the antennas and their height, the expected receiving quality, the propagation path of the signals, the geographical location and terrain, the presence of interference, the receiver sensitivity and transmitter power. Given a receiver and a fixed location the adjustable parameters are the antennas and their height and the transmitter power. In this case, ITU-R P.1546-1 provides easy-to-follow procedures to calculate the field strength given the antenna height and transmitter power [85].

ITU-R P.1546-1 is the ITU Recommendation for point-to-area field strength predictions for terrestrial services in the frequency range 30 MHz to 3000 MHz. Land use only is considered. Based on the recommendation ITU-R P.1546-1, a simulation is built up in the following way:

Step 1: The dimensionless parameter k is calculated using the transmitter or repeater height h, as follows:

$$k = \frac{log[\frac{h}{9.375}]}{log(2)} \tag{8.4}$$

Where h is in the range of 9.375 and 1200m; k is an integer in the range between 0 and 7.

Step 2: An intermediate field strength E_u at the distance d for transmitter height h is calculated as follows:

$$E_u = p_b \times log \frac{10^{\frac{E1+E2}{p_b}}}{10^{\frac{E1}{p_b}} + 10^{\frac{E2}{p_b}}} \tag{8.5}$$

Where

$$p_b = d_0 + d_1 \cdot \sqrt{k} \tag{8.6}$$

$$E1 = (a_0 \cdot k^2 + a_1 \cdot k + a_2) \cdot log(d) + 0.1995 \cdot k^2 + 1.8671 \cdot k + a_3 \tag{8.7}$$

$$E_2 = E_{off} + E_{ref} \tag{8.8}$$

Where

$$E_{off} = \frac{C_0}{2} \cdot k \cdot k[1 - tgh[c_1 \cdot [log(d) - c_2 - \frac{c_3^k}{c_4}]]] + c_5 \cdot k^{c_6} \tag{8.9}$$

$$E_{ref} = b_0[exp[-b_4 \cdot 10^{\xi}] - 1] + b_1 \cdot exp[-(\frac{log(d) - b_2}{b_3})^2] - b_6 \cdot log(d) + b_7 \tag{8.10}$$

Where

$$\xi = log(d)^{b_s} \tag{8.11}$$

In the equations in step 2 above a_0 to a_3, b_0 to b_7, c_0 to c_6, and d_0 to d_1 are parameters given in Table 8.2. Because DVB-H is most likely to be used in UHF band (470-838MHz) and L band (1440-1790MHz) [126], only the transmitting frequencies 600MHz (in UHF band) and 2000MHz (adjacent to L band for convenience) are used and different time percentages (50%, 10% and 1%) for land area are used.

Step 3: The final field strength E_b at the distance d for transmitter height h is:

$$E_b = p_b b \cdot log[\frac{10^{\frac{E_u+E_{fs}}{p_{bb}}}}{10^{\frac{E_u}{p_{bb}}} + 10^{\frac{E_{fs}}{p_{bb}}}}] \tag{8.12}$$

In the above equation E_{fs} is the free space field strength assuming that the transmitter $E.R.P.$ (Effective Radiated Power)is 1KW and E_{fs} is given by:

$$E_f s = 106.9 - 20log(d) \qquad dB(\mu V/m) \tag{8.13}$$

And P_{bb} in equation (7.12) is the blend coefficient set to value 8 according to [56].

Table 8.2. Coefficients for the Generation of the Land Tabulations

Frequency	600MHz			2000MHz		
Time(%)	50	10	1	50	10	1
a_0	0.0946	0.0913	0.0870	0.0946	0.0941	0.0918
a_1	0.8849	0.8539	0.8141	0.8849	0.8805	0.8584
a_2	-35.399	-34.160	-32.567	-35.399	-35.222	-34.337
a_3	92.778	92.778	92.778	94.493	94.493	94.493
b_0	51.6386	35.3453	36.8836	30.0051	25.0641	31.3878
b_1	10.9877	15.7595	13.8843	15.4202	22.1011	15.6683
b_2	2.2113	2.2252	2.3469	2.2978	2.3183	2.3941
b_3	0.5384	0.5285	0.5246	0.4971	0.5636	0.5633
b_4	4.323×10^{-6}	1.704×10^{-7}	5.169×10^{-7}	1.677×10^{-7}	3.126×10^{-8}	1.439×10^{-7}
b_5	1.52	1.76	1.69	1.762	1.86	1.77
b_6	49.52	49.06	46.5	55.21	54.39	49.18
b_7	97.28	98.93	101.59	101.89	101.39	100.39
c_0	6.4701	5.8636	4.7453	6.9657	6.5809	6.0398
c_1	2.9820	3.0122	2.9581	3.6532	3.547	2.5951
c_2	1.7604	1.7335	1.9286	1.7658	1.7750	1.9153
c_3	1.7508	1.7452	1.7378	1.6268	1.7321	1.6542
c_4	198.33	216.91	247.68	114.39	219.54	186.67
c_5	0.1432	0.1690	0.1842	0.1309	0.1704	0.1019
c_6	2.2690	2.1985	2.0873	2.3286	2.1977	2.3954
d_0	5	5	8	8	8	8
d_1	1.2	1.2	0	0	0	0

The relationship between the receiver-received quality of service Q and the consumed battery power C can be analytically expressed as equation (8.14) and equation (8.15) below:

$$Q = \alpha \cdot f(h_1, h_2, h_3, l_1, l_2, l_3) \tag{8.14}$$

$$C = \beta \cdot g(l_1, l_2, l_3) \tag{8.15}$$

Where $Q = \alpha \cdot f(h_1, h_2, h_3, l_1, l_2, l_3)$ and $C = \beta \cdot g(l_1, l_2, l_3)$ are the abstracted functions obtained from equations (8.2) and (8.3); while l_1, l_2 and l_3 are the corresponding distances from the DVB-H receiver to the central main transmitter and the nearest two repeaters and h_1, h_2 and h_3 are the corresponding antenna height of the central main transmitter and the nearest two repeaters.

Based on the field strength prediction procedures above a simulation model is built in MATLAB. The simulation parameters are: DVB-H cell radius is 30km; antenna height is between 9.375 and 1200m; 600MHz, land path, 50% time. After simulation, the relationship obtained between E_b and A_r is shown in Fig. 8.5. In Fig. 8.5 h_i(i= $1, 2, \cdots, 5$) is the antenna height of the main transmitter and the repeaters.

For simplicity it is supposed that α and β are both equal to 1 then the receiver-received quality of service Q and the consumed battery power C is shown in Fig. 8.6. It is easy to see that the received quality of service Q is increased with the increasing of the battery power consumption C and given a fixed value of cell radius, antenna height, transmitter power, transmission frequency and time percent a fixed relationship between Q and C is able to be obtained.

Now the cost of the RA_handover scheme is considered. Active repeaters are expensive. On the other hand, the cost of the repeaters is connected with the cost of the main transmitter. Because low power repeaters cover small areas, in order to provide the same quality of service for the users even in the border area of the cell it is necessary to install high power main transmitters that will be very costly. Without repeaters the main transmitter must use high power to provide acceptable quality of service in the cell border area. The more transmission power the main transmitters and repeaters have, the more costly they will be. However, it is not very easy to get the exact cost of installing the repeaters and the main transmitters. Though it is hard to compare the exact cost of the RA_handover algorithm and the algorithm without the repeaters, it can be seen that by implementing the RA_handover algorithm the handheld DVB-H receiver can save considerable battery power consumption and improve the quality of service. This will drive the consumers' desire to use the DVB-H service, which is equates to profit making for the whole system. The exact cost comparison will be done in the future.

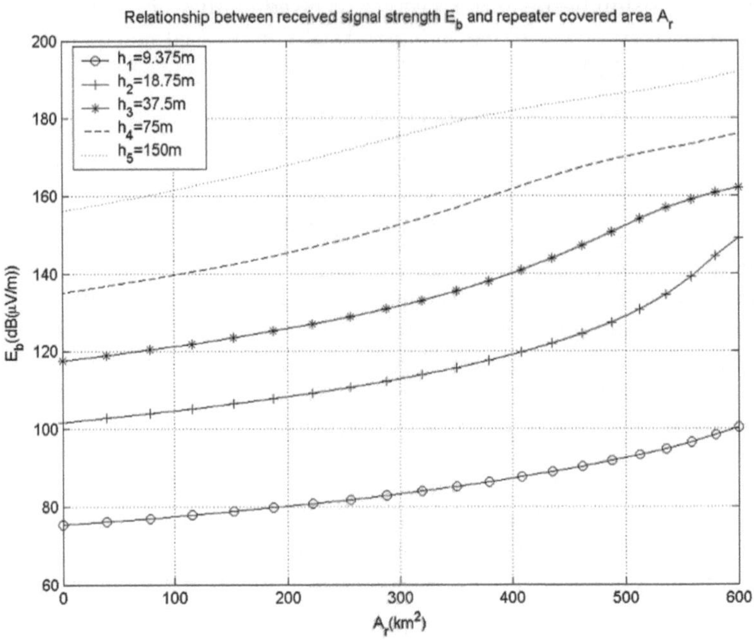

Fig. 8.5. The Relationship Between Received Signal Strength and Repeater Covered Area

8.5 Conclusions

Handover for unidirectional broadcasting networks like DVB-H is a novel issue and a new challenge. Low power transmitters constituting dense multi-frequency cellular DVB-H networks could make handover a very important issue in DVB-H network planning and optimisation. This chapter has described a novel approach for DVB-H receivers with MIMO antennas support to decide when soft handover should occur, called RA_handover, based on a proposed intelligent repeater structure. A simple mathematical calculation showed that the RA_handover scheme could save at least 25% of the battery power consumed by the handover decision-making algorithm compared with that of handover algorithm without repeaters. A simulation model has also been developed to show the performance of the RA_handover approach. Simulation results showed that the receiver-received quality of service is increased with the increase of the repeater-covered area. And the maximum quality of service happens when the receiver always feels it is located in the centre of the cell. The cost issues introduced by the RA_handover algorithm are also

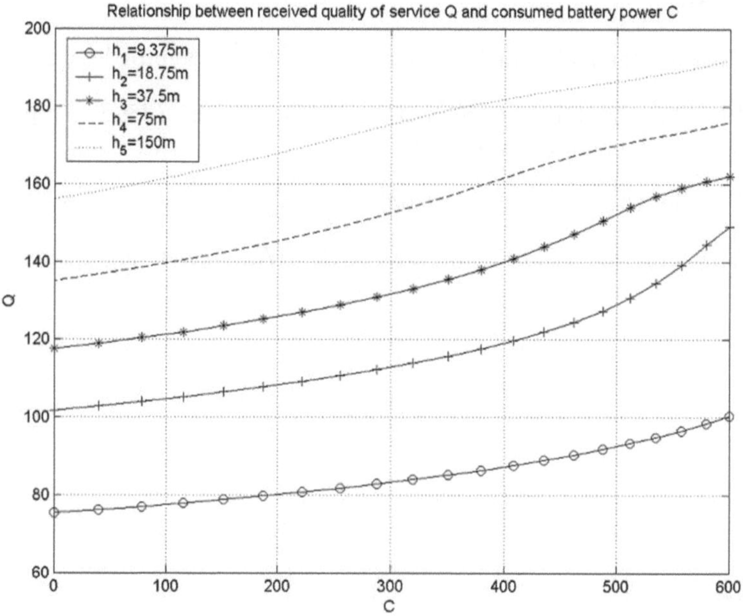

Fig. 8.6. The Relationship Between Received Quality of Service and Consumed Battery Power

analysed. Although it is still difficult to get the exact amount of the cost that would be incurred by introducing the RA_handover algorithm and the comparison of cost with the handover algorithm without repeaters is not very easy. It has already been shown that the cost will not be an obstacle for the implementation of the RA_handover algorithm when the overall system costs and revenue are considered.

In the RA_handover algorithm, the repeaters are active repeaters. These repeaters can improve the quality of the services in the repeater-covered area as described in this chapter. The active repeaters can also add extra signalling information to the received signals. For the service providers, the extra signalling information can be used to signal the services or even the additional localized services in the repeater covered area. This will definitely provide an extra tool for the management of the provided services. Since the RA_handover is a very feasible handover algorithm of DVB-H, as demonstrated through simulation results reported in this chapter, it is very promising to be considered in the standardization process and to be eventually incorporated into the soft handover standard for DVB-H.

Problems

8.1. What are the main functions of the repeaters in the broadcast networks?

8.2. What are the different kinds of repeaters and which kind of repeater is used in the RA_handover algorithm?

8.3. What is RA_handover algorithm?

8.4. What are the possible drawbacks of the RA_handover algorithm?

9

Repeater Aided Soft Handover Probability

For network planning and optimisation purposes this Chapter develops a mathematical model for calculating the soft handover probability of the RA_handover algorithm described in Chapter 8. To give an indication of the order of magnitude of the percentage of the power consumed by the handover decision-making algorithm that could be saved, the mathematical model shows that RA_handover can reduce average front-end power consumption by up to 63.22% in the worst case compared with the handover method described in [104].

The handover process in DVB-H will be initialized and finished by the handheld device alone. In DVB-H networks repeaters were originally used to extend the coverage area and cover shadow areas such as tunnels, deep valleys, subterranean locations and indoor areas etc. By using repeaters low power transmitters are possible. Low power transmitters constitute dense multi-frequency cellular structured DVB-H networks that can provide more service capacity.

In the RA_handover scheme, a novel low power intelligent repeater structure was proposed in Chapter 8. Each repeater-covered area is defined as one subcell. Unlike a passive repeater that simply amplifies and relays an incoming signal, an intelligent repeater can demodulate the incoming transport stream, and remodulate the transport stream with handover scheme information and subcell id information in the PSI/SI table and the TPS bits. Specifically the subcell frequency information is located in the cell_frequency_link_descriptor and the frequency_list_descriptor; the subcell coverage information is located in the cell_list_descriptor; the subcell identification information is located in the TPS bits. After remodulation the repeater sends the signal to the transmitting antennas. The structure of such an intelligent repeater is shown in Fig. 8.2.

In Fig. 8.2, the TPS Adapter adds the unique repeater specific information to PST/SI table and TPS bits in the transport stream.

For the proposed RA_handover scheme, intelligent repeaters are put uniformly around the cell borders in a dense multi-frequency cellular structured

DVB-H network. Each Repeater covered area is one subcell. The DVB-H mobile terminal will have MIMO (Multi-Input Multi-Output) [11] antenna systems installed. One of the advantages of a MIMO system is that the terminal receiver can distinguish the different transport streams from the main transmitter and repeater even if they use the same frequency. When a mobile device moves into such a subcell it receives the unique repeater identification information from the repeater transmitted signals. Since the repeaters have unique identification information, the mobile device will know in which specific subcell it is located. When the device is in more than one such subcell the device then begins to measure the SNR value from the two or three different subcells in which it is located. Once the SNR threshold margin value s_th is reached for a certain threshold time t_th, the receiver will tune to the frequency with the strongest SNR value to continue service reception. One example of the SNR and the duration threshold is shown in Fig. 2.4. The RA_Handover algorithm is further illustrated in Fig. 9.1.

9.1 Network Topology for Handover probability

Fig. 9.2 and Fig. 9.3 further illustrate the RA_handover algorithm. The SNR (Signal to Noise Ratio) in Fig. 9.3 is used to refer to the signal strength that the receiver needs to measure in order to decide the target handover cell.

Fig. 9.2 shows a seven cell DVB-H network topology. Each cell contains six repeaters, i.e. six subcells, around the border of the cell. R_{12} is one subcell in cell 1 near the border between cell 1 and cell 2. R_{21} is one subcell in cell 2 near the border between cell 2 and cell 1. R_{13} and R_{31} are the corresponding subcells bordering cell 1 and cell 3 respectively.

Suppose the whole network is broadcasting the same service, so the terminal does not need to seek an alternative service (which could be a local variation of the original service or an associated service) when it moves around in the network, that means that the terminal need not check the service_list_descriptor of each transport stream to find the service_id of the previously selected service. The center frequency used in the cell is given in the terrestrial_delivery_system_descriptor, while for every other cell the frequency is given in the frequency_list_descriptor. When the mobile terminal moves into the R_{12} area, it will decode the subcell id information from the frequency_list_descriptor and the cell_list_descriptor that corresponds to the corresponding repeater covered area. This process happens in the burst time period. At this location the receiver begins to carry out handover measurement during the off burst period. When the received signal strength from cell 2 meets the requirement for both s_th and t_th the terminal will decode the transport stream from cell 2 and check for the same pair of original_network_id and transport_stream_id to handover to the target cell 2. Otherwise, it does not handover. When the terminal is in the comparatively small shaded area that borders cell 1, cell 2 and cell 3, it decodes its location information from

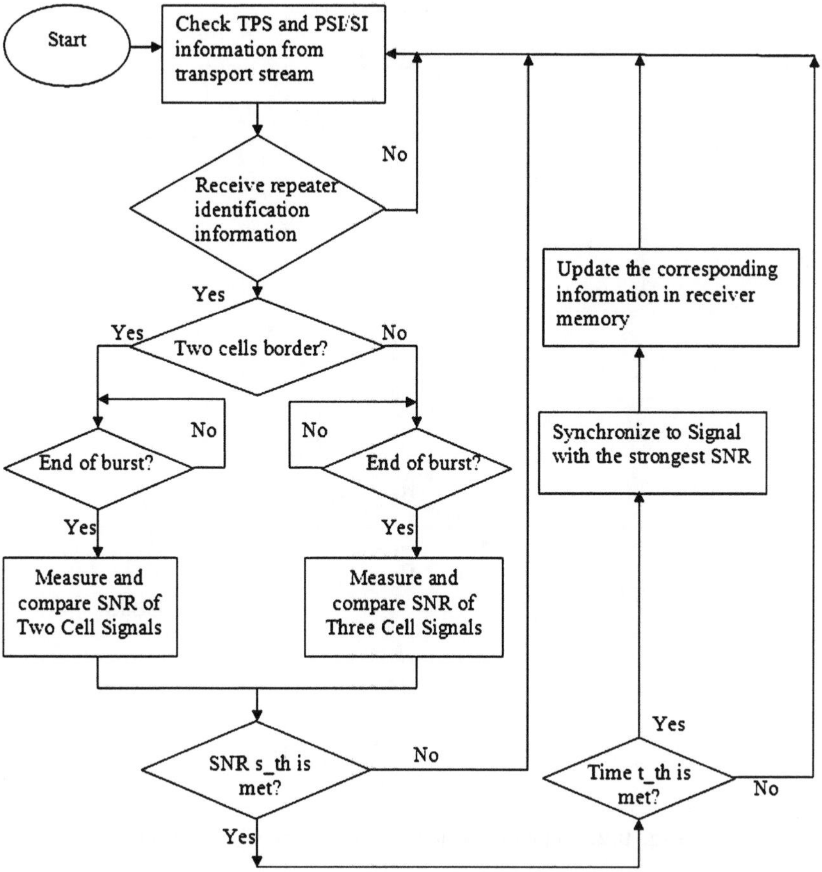

Fig. 9.1. Repeater Aided Soft Handover Algorithm

the frequency_list_descriptor and the cell_list_descriptor and then will measure signals from the three different subcells in the off burst time. The handover decision is made after the signal strength measurement.

Since all the subcells have the same frequency as their master cells no handover will happen between the master cell and the subcells. Here master cell refers to the main transmitter covered cell area. Because the original signals are usually very weak in the cell border area and a MIMO antenna is used in the terminal, the interference between the original signals and the repeater-modulated signals can be neglected. There are already some Papers about this kind of co-channel interference in MIMO systems [15, 12, 13]. These show that installing intelligent repeaters does not ruin the SFN (Single Frequency Network) performance. Details of these kinds of interference study in DVB-H

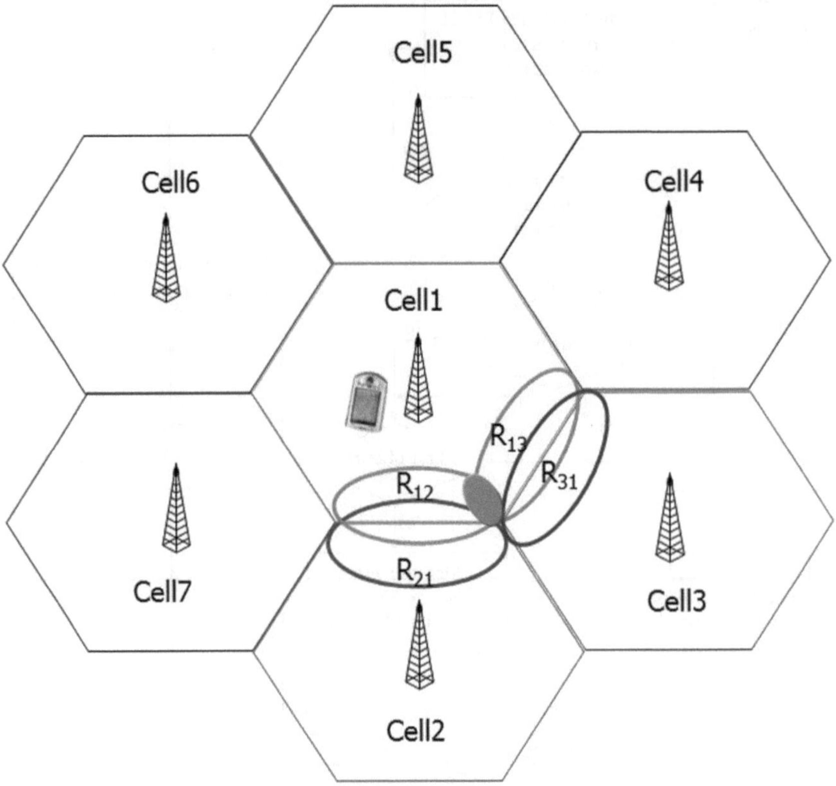

Fig. 9.2. Repeater Aided Soft Handover Cell Structure

will be presented in later research. In this way the receiver does not need to measure the signal strength level constantly, thus saving battery power. With optimized repeater location planning and the use of directional MIMO antennas, frequent unnecessary handovers can be reduced in the subcells

From the handover probability point of view, if the size of the subcell is decreased the handover probability is decreased, too. With decreasing handover probability, more battery power can be saved. However, if the subcell area is continuing to shrink, the terminal may continue to expect the repeater signals thus missing the best moment to carry out the handover. In this case, the service robustness will also be decreased to some extent. So the trade-off between handover probability and service robustness/quality of service must be optimized in the network planning stage.

9.2 Mathematical Model for Reduced Power Consumption

Radio network planning is responsible for proper handover parameter setting and site planning so that the soft handover probability does not exceed the desired value. Typically, the soft handover probability is required to be kept below a certain limit so that the receiver can receive reliable services and save battery power. In this section a mathematical model is developed to calculate the average soft handover probability in a cell for the given threshold s_th and t_th. This probability can be calculated by taking the ratio of the surface area of the part of the network where soft handovers are possible, relative to the total network surface area [39].

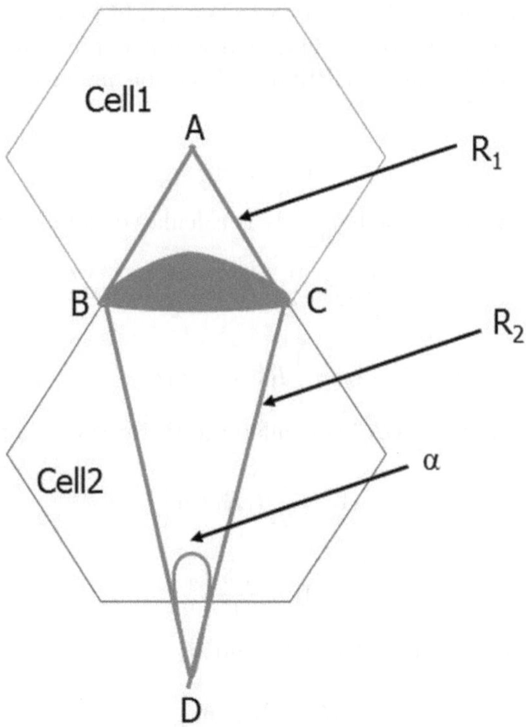

Fig. 9.3. Calculation of Soft Handover Probability

Fig. 9.3 is used to calculate the soft handover probability SH_PROB. In Fig. 9.3, A_t is the area of triangle ABC, A_s is the shaded area. R_1 is the radius of cell 1. For the convenience of calculation, let R_2 denote the radius of the circle that creates the shaded area and a denote the angle that creates the shaded area. Both α and R_2 are decided by the repeater's location and antenna pattern.

In Fig. 9.3, it is easy to see that the smaller the shaded area A_s, the smaller the soft handover probability SH_PROB. But if A_s is too small, soft handover will not happen as expected, thus service interruption will happen when the mobile receiver moves from one cell to another. If As is too big, the receiver will spend more time measuring the signal SNR value, thus wasting a lot of battery power. The optimum trade off between received service quality and consumed battery power has to be well planned. The trade off depends on the cell size, context environment, user behavior, etc.

For example, in densely populated urban areas the moving speed of the users is slow and users tend to be static most of the time. In this case, the area As could be small. On the other hand, on a high way in a rural area, the moving speed of the users is normally fast and users tend to be moving most of the time. In this case, the area As should be large to avoid service interruption when users are moving across cell borders at high speed. For simplicity in the following SH_PROB calculation process radio propagation effects are neglected.

Clearly,

$$0 < \alpha < \pi \tag{9.1}$$

Using Fig. 9.3, the shaded area A_s is calculated as follows,

$$A_s = \frac{\alpha}{2}R_2^2 - \frac{1}{2}(2R_2 \sin\frac{\alpha}{2})R_2 \cos\frac{\alpha}{2} \tag{9.2}$$

$$A_s = \frac{1}{2}R_2^2(\alpha - \sin\alpha) \tag{9.3}$$

Because the triangle ABC is equilateral, the area of the triangle ABC is,

$$A_t = \frac{1}{2}R_1 R_1 \sin\frac{\pi}{3} \tag{9.4}$$

$$A_t = \frac{\sqrt{3}}{4}R_1^2 \tag{9.5}$$

From the relationship between R_1 and R_2,

$$R_1 = 2R_2 \sin\frac{\alpha}{2} \tag{9.6}$$

The area of the triangle ABC expressed using R_2 and α is obtained as,

$$A_t = \frac{\sqrt{3}}{4}(2R_2 \sin\frac{\alpha}{2})^2 = \frac{\sqrt{3}}{2}R_2^2(1 - \cos\alpha) \tag{9.7}$$

Thus $\frac{A_s}{A_t}$ only depends on α,

$$\frac{A_s}{A_t} = \frac{\alpha - \sin\alpha}{\sqrt{3}(1 - \cos\alpha)} \tag{9.8}$$

The plot of $\frac{A_s}{A_t}$ as against α is shown in Fig. 9.4.

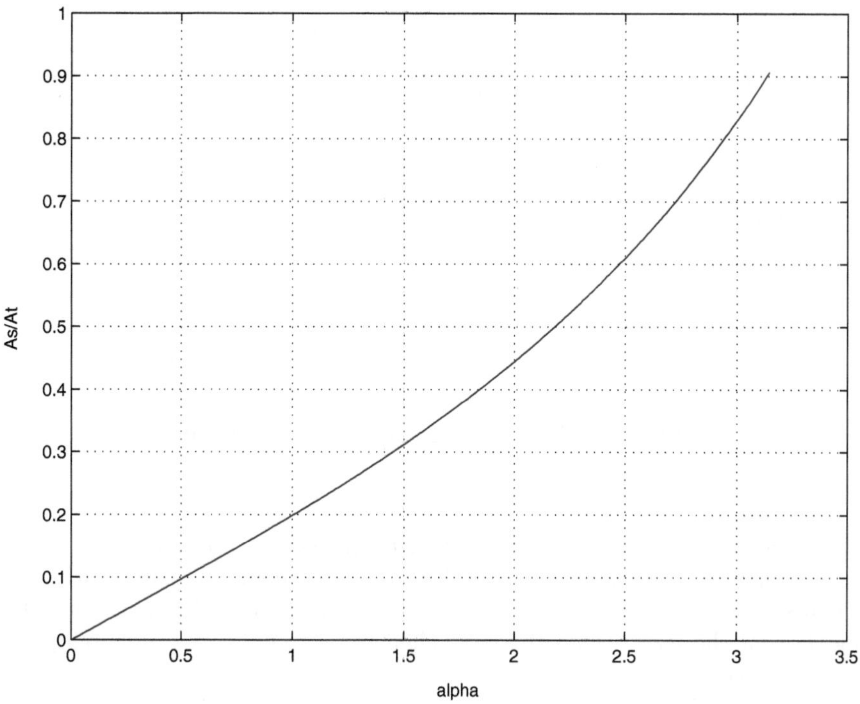

Fig. 9.4. $\frac{A_s}{A_t}$ as a function of α

The next task is to examine whether $\frac{A_s}{A_t}$ as a function of α is normally distributed or not. A normal probability distribution was fitted to $\frac{A_s}{A_t}$ using MATLAB and the result is shown in Fig. 9.5.

From Fig. 9.5 it is easy to see that $\frac{A_s}{A_t}$ as a function of α is well approximated by a normal distribution.

Suppose the probabilities P_s and P_t of a mobile device satisfying the s_th and t_th requirements are both normally distributed without loss of generality.

$$P_s = \frac{1}{\sqrt{2\pi\sigma_s^2}} e^{-\frac{(x-\mu_s)^2}{2\sigma_s^2}} \tag{9.9}$$

$$P_t = \frac{1}{\sqrt{2\pi\sigma_t^2}} e^{-\frac{(x-\mu_t)^2}{2\sigma_t^2}} \tag{9.10}$$

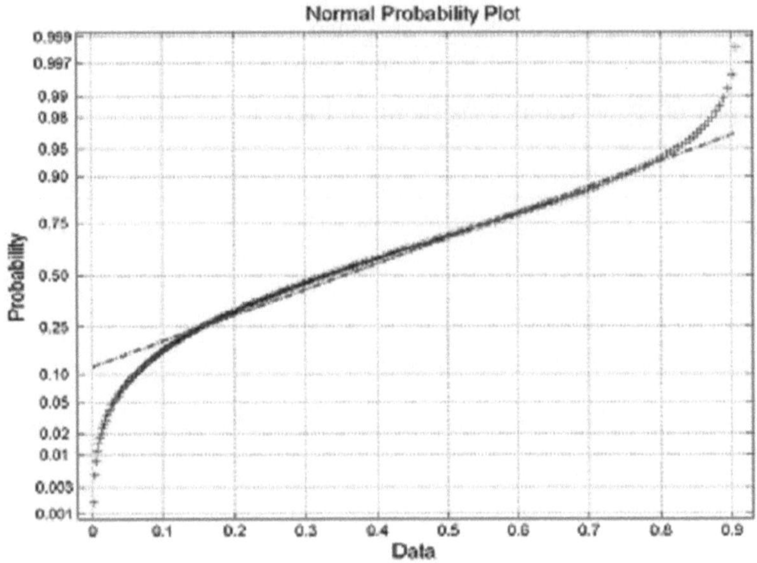

Fig. 9.5. $\frac{A_s}{A_t}$ plotted against Normal Probability

To get an indication of the order of magnitude of the percentage of the power that would be saved by using the proposed scheme compared with using the first handover scheme it is assumed that P_s and P_t are independent of the location of the user within the cell. Then the soft handover probability can be expressed as:

$$SH\_PROB = \frac{A_s}{A_t} P_s P_t = \frac{\alpha - \sin\alpha}{\sqrt{3}(1 - \cos\alpha)} P_s P_t \tag{9.11}$$

In this case, SH_PROB is normally distributed.

Suppose the mean values of P_s and P_t are E_s and E_t respectively. From Fig. 9.4 the mean handover probability for RA_handover algorithm is obtained as:

$$mean(SH\_PROB) = 0.3678 E_s E_t \tag{9.12}$$

For the proposed soft handover algorithm in [104], because the receiver measures the signal strength without prediction its soft handover probability will be $P_s P_t$. in the worst case when the terminal uses every off burst time to carry out handover measurement. This means that on average the RA_handover algorithm would save around two thirds of the battery power usage compared with the one proposed in [104].

9.3 Conclusions

Handover for unidirectional broadcasting networks is a novel issue. Low power transmitters constituting dense multi-frequency cellular structured DVB-H networks would make handover a very important issue in DVB-H network planning and optimization. By using the Repeaters the additional cost imposed by the GPS receiver on the DVB-H handheld terminals is transferred to network side which is more economic from the marketing point of view. In other words, since the number of terminals is far higher than the number of repeaters, the implementation cost on the network side is typically less critical compared to the terminal side. The RA_handover algorithm also reduces the *Ping Pong* effect and the cause of "fake signals" while improving the quality of service at the same time. A Mathematical model has been presented for calculating the soft handover probability of the algorithm. Since lower probability of handover means lower power consumption, to get an indication of the order of magnitude of the percentage of the power consumption that could be saved using the RA_handover algorithm instead of the proposed handover method in [104], analysis has been provided showing that as much as 63.22% of the front-end battery power may be saved compared with the one in [104]. The drawback of the RA_handover algorithm is that it will bring the extra cost to the network operator in the early stage of DVB-H deployment.

Problems

9.1. How is the soft handover probability being calculated?

10

Handover Algorithm for Converged Networks

DVB-H is not only a broadcast standard. It also defines the use of other technology as an interactive uplink. In order to provide interactive services for DVB-H, UMTS can be used as a terrestrial interaction channel for the unidirectional DVB-H network. The converged DVB-H and UMTS network can be used to address the congestion problems due to the limited multimedia channel access in the UMTS network. In the converged network, intersystem soft handover between DVB-H and UMTS is needed for optimum radio resource allocation and to reduce network operation cost while providing the required quality of service. This chapter deals with the intersystem soft handover between DVB-H and UMTS in such a converged network. The converged network structure is presented. A throughput based soft handover scheme is proposed and evaluated. After considering network operation cost, the performance tradeoff between network quality of service and network operation cost for intersystem soft handover in the converged network is modeled using a stochastic tree and analyzed using numerical simulation. The results show that the proposed algorithm is feasible and has the potential to be used for implementation in the real environment [27].

10.1 Introduction

DVB-H targets mainly IP data services, called IP Datacast. It is assumed that a large number of the IP Datacast services will be of interactive nature [152]. A DVB-H network is a unidirectional broadcasting network that provides a means to deliver large quantities of popular content to large user groups at relatively low cost but is very limited in providing interactivity and personalization of content. A UMTS network, as a cellular telecommunications network, does not provide a low cost delivery mechanism for downloading large quantities of data simultaneously to large user groups but has the excellent properties of interactivity and personalization. UMTS is therefore an excellent system to support IP Datacast by providing a terrestrial interaction channel for DVB-H. In such a converged network, DVB-H works as a high bit rate download channel, while UMTS provides the interaction channel for DVB-H. Another application area of such a converged network is the so-called "hot spot" area where the UMTS network cannot provide the requested data delivery services because of its limited data channels. In this case, DVB-H can provide an alternative way to deliver the requested data delivery services to relieve the congested UMTS cells.

DVB-H will broadcast the same data services to as many users as possible within its coverage area due to its broadcast nature. When only a few users request the same service, using DVB-H as a downlink channel will waste bandwidth that could be utilized by many more users. In this case, UMTS is an efficient way to provide the service to the limited number of users. On the other hand, when the number of users requiring the same service exceeds a certain limit, DVB-H will be a profitable way to provide the downlink channel. The intersystem handover in this chapter is not defined as service continuation when user moves out of one network covered area but as the optimum radio resource reallocation between UMTS and DVB-H when the user is in an area covered by both UMTS and DVB-H networks. The scenario of the users moving out of the range of one technology will not be discussed in this chapter. The service of the converged network referred to in this chapter is IP data delivery as it is a service that can exploit intersystem handover in an easily described way. Because the handover in DVB-H is always soft handover and because of the better performance of soft handover compared with hard handover in service continuation, this chapter focuses on the intersystem soft handover in the converged network.

So far, intersystem handover studies for UMTS have only focused on handover between UMTS and GSM [118, 119] and between UMTS and WLAN [120, 121]. No literature is available concerning handover between UMTS and DVB-H in a converged DVB-H/UMTS network at the time of writing this chapter. In this chapter, an intersystem soft handover scheme is presented using the throughput based load factor as the handover criteria for the converged network and evaluated using a Markov chain model. Since network operation cost is also an issue in the converged network, a stochastic tree model is used to analyze the handover scheme when considering both network operation cost and network quality of service. The rest of the chapter is organized as follows:

Section 10.2 provides the background of the research work in the intersystem handover domain. Section 10.3 provides an overview of the proposed converged network structure and of the intersystem handover issue. Section 10.4 presents the intersystem soft handover algorithm between UMTS and DVB-H, including the handover measurement, handover criteria, handover execution and evaluation. In Section 10.5, the stochastic tree concept is introduced into the communication area and a stochastic tree model is set up to analyze the handover scheme and a numerical simulation is presented. The optimum tradeoff between the converged network operation cost and quality of service is obtained from simulation results. Section 10.6 concludes the chapter.

10.2 Research Background

Although there are no previous concrete works regarding the convergence of DVB-H and UMTS, some relevant discussions have already been made in various literatures.

Paper [24] presents a signalling protocol for the exchange of information between a network management system and intelligent multimodal wireless terminals in a composite radio networks where different radio networks are available, and some preliminary measurement work were done mainly between WLAN, GPRS and DVB-T. Although the idea of redistributing terminals over different radio networks was pointed out in [24], the description was very simple and it did not present how such redistribution would be performed in detail steps.

Paper [35] presents two typical heterogeneous wireless network scenarios with DVB-T, WLAN and GPRS. The first scenario refers to the redistribution of traffic between DVB-T and the other available access technologies. The second scenario refers to the users moving out of the multi-coverage regions and having only DVB-T access available. However, it lacks the detail procedures how the redistribution in the first scenario is triggered and performed.

Paper [34] focuses on the integration of WLAN and GPRS by describing how the two networks are interworking each other mainly from the architecture point of view while omitting the radio resource optimization part. Paper [33] proposed a way for the UMTS network operators to forecast the traffic of the high-demand services and transfer them to a cooperating DVB-T network when appropriate. However no detail procedures of transferring DVB-T traffic to UMTS is described.

While the papers [118] [119] [120] [121] are focusing on how the intersystem handover are actually performed between either UMTS and GSM or UMTS and WLAN, the papers [24] [35] [34] [33] are focusing on the radio resource optimization between the broadcast and the telecommunications networks. Besides, all the above mentioned works did not consider the trade-off between the network operation cost and the network quality of service and no quantization of such parameters are available.

Based on the thoughts of the previous works, this chapter tries to combine both of the above two ways of thinking by proposing a novel intersystem soft handover algorithm between DVB-H and UMTS. While this chapter describes the way how the intersystem handover should be performed on the terminal part as those briefly mentioned in chapter 6, more spaces of this chapter are given to the radio resource optimization part of the intersystem handover algorithm. While the compressed mode in UMTS and the time slicing mode in DVB-H are utilized for handover process, the load factor is proposed as the threshold based approach criteria and a novel stochastic tree algorithm is proposed for the optimum radio resource allocation part in a proposed clear and practical converged network architecture. As an obvious evolution of the previous work, the network operation cost and network quality of services are quantized and simulated. For simplicity and focus reasons, the service continuation scenario in the converged networks caused by the users moving out of one network coverage area is not considered in this chapter.

10.3 Converged Network Overview

Multi-tier networks that provide coverage to the same areas using several cells of different sizes are useful to accommodate high traffic density while providing high quality of service [16]. The proposed converged DVB-H and UMTS network cell structure is a hierarchical cell structure in which the DVB-H cells overlay the UMTS cells. Therefore, the converged network can be regarded as a multi-tier network from the cell structure point of view, as shown in Fig. 10.1

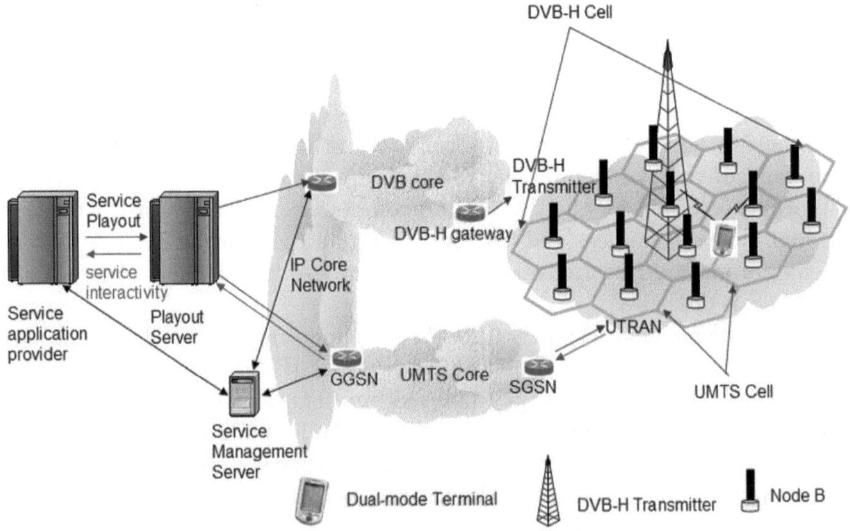

Fig. 10.1. Converged DVB-H and UMTS network structure

In the converged DVB-H and UMTS network, the DVB-H cells cover the same area as the UMTS cells do. Because the DVB-H cell size is usually larger than the UMTS cell size, one DVB-H cell normally covers several UMTS cells. Hence, the converged network has a two-layer macro-micro cell structure, in which the DVB-H cell is in the macro layer while the UMTS cell is in the micro layer. This overlapping cell structure of the converged network is very useful in a "hot spot" area that is defined as an area where traffic load is substantially larger than the designed load [17], such as sports arena, concerts stadium, and theater, etc.

In order for the terminal to be able to perform handover between DVB-H and UMTS, either the terminal has two independent receivers or the terminal has a single dual mode receiver. For handheld devices, it will reduce complexity and cost to have a single dual mode receiver rather than have two independent receivers. The UE (User Equipment) referred to in this chapter is assumed to be such a dual mode handheld terminal.

In Fig. 10.1, the downlink stream from the Playout Server has two paths to reach the UE. One is through the DVB-H network; the other one is through the downlink channels in UMTS. The uplink stream packets from the UE have only one path that is the UMTS uplink channel.

In the converged network, the DVB-H network part and the UMTS network part have a common core network that is the IP network. It is important to note that the data in DVB-H network are IP packets not MPEG2 packets like that of DVB-T. There is also a SMS server (Service Management Server)located in the IP core network. The SMS server is the key component in charge of the radio resource allocation and intersystem soft handover between DVB-H and UMTS.

Handover within the converged network takes three forms: handover within the DVB-H network, handover within the UMTS network and intersystem handover between the DVB-H network and the UMTS network. This chapter only considers the intersystem handover between DVB-H and UMTS.

The intersystem handover between UMTS and DVB-H in the converged network is different from the intersystem handover between UMTS and other radio access networks like GSM. In the DVB-H/UMTS converged network there is always an uplink channel from the User Equipment (UE)to the IP core network through the UMTS network. Therefore, the handover between UMTS and DVB-H only refers to a downlink channels handover that should have no significant effects on the uplink channels of the UE. In this intersystem soft handover scheme, the downlink channel in UMTS only refers to the dedicated channel (DCH)because only the DCH can support soft handover in UMTS. Another difference from the other intersystem handover schemes is that the intersystem handover in a DVB-H/UMTS converged network is triggered not only by technical issues like physical constraints and better radio resource allocation but also by economic issues like network operation cost. The network costs issue is important for intersystem handover in a converged DVB-H/UMTS network because the converged network may be jointly operated by a broadcasting network operator and a telecommunications network operator.

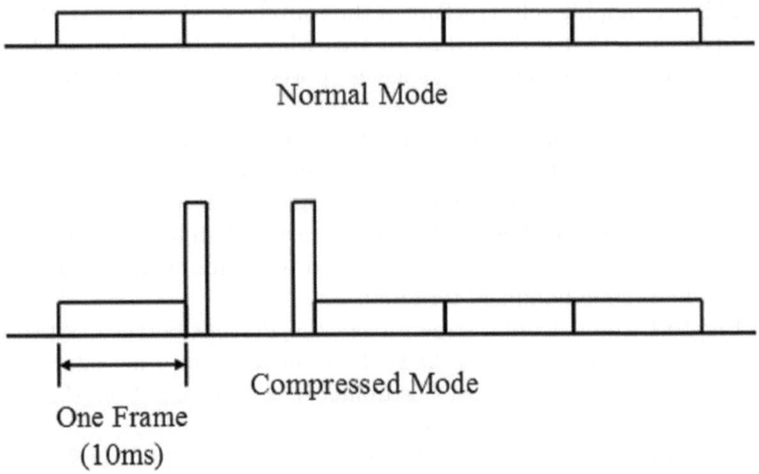

Fig. 10.2. Comparison between Normal Mode and Compressed Mode

10.4 Handover Between UMTS and DVB-H

The intersystem soft handover mechanism between UMTS and DVB-H in the converged network can be divided into the following three stages:

- Handover measurement
- Handover decision based on the handover criteria
- Handover execution

10.4.1 Performing DVB-H Measurements with the Compressed Mode of UMTS

UMTS has two modes: Frequency Division Duplex (FDD) mode and Time Division Duplex (TDD) mode. For simplicity, only the FDD mode is considered in this chapter. In FDD mode, the UE is continuously receiving on the UMTS downlink channel. This normal continuous mode is not suitable for intersystem measurement that needs transmission gaps. The compressed mode introduced in the 3GPP standard [84] has therefore been proposed to perform the intersystem handover between UMTS and GSM [118, 119].

UMTS transmits data using 10 milliseconds long frames. Each frame consists of 15 time slots. In the compressed mode, some frames are compressed. In the compressed frames, some slots are not used for transmitting data thus creating transmission gaps. The instantaneous transmit power is increased in the compressed frame in order to keep the quality, Bit Error Rate (BER),

Frame Error Rate (FER), etc., unaffected by the reduced processing gain. The compressed mode and the normal mode frame structures are compared in Fig. 10.2. In compressed mode the first few milliseconds and the last few milliseconds in a frame are used to transmit all the information in the frame, and the remaining duration of the frame is used to carry out intersystem soft handover measurement. In this way, when performing soft handover from UMTS to DVB-H, the compressed mode is used to create transmission gaps as measurement periods for the terminal to make the required measurements of DVB-H network parameters.

10.4.2 Performing UMTS Measurements with the Time Slicing Mode of DVB-H

The time slicing mode was introduced in the DVB-H standard [67] to reduce the average power consumption of the terminal and enable smooth seamless soft handover. In time slicing mode, the data packets are sent in periodic bursts using significantly higher instantaneous bit rates compared to the bit rates required if the data were transmitted using continuous mode like that of DVB-T [70]. The continuous transmission mode and the time slicing transmission mode are illustrated in Fig. 10.3. When performing handover from DVB-H to UMTS, the UE carries out handover measurement of UMTS network parameters using the off burst time of the time slicing mode. Because the intersystem handover measurement always happens in the transmission gaps for both UMTS and DVB-H, the handover between the two networks is soft handover and can be seamless.

10.4.3 Intersystem Handover Criteria

The intersystem handover between UMTS and DVB-H in the converged network is both a technical issue and an economic issue and this must be accounted for in choosing the handover criteria.

In this section, only the technical issue is considered which is making the best radio resource allocation according to the network quality of service. The economic issue is discussed later in this chapter using stochastic tree models.

The UMTS network plays a key role in the intersystem soft handover process because of its interactive nature. In order to fully utilize the advantages of the DVB-H network, the UMTS network loading status and the DVB-H network utilization status are chosen as the intersystem soft handover criteria. Such intersystem soft handover criteria are considered only from the network side here. For the implementation of soft handover in the terminal, signal strength or SNR (Signal Noise Ratio) has also to be used as a handover criterion [104] [106].

Two different approaches can be taken to measure the load of the UMTS air interface. The first defines the load via the received and transmitted wideband power; the second is based on the sum of the bit rates allocated to all

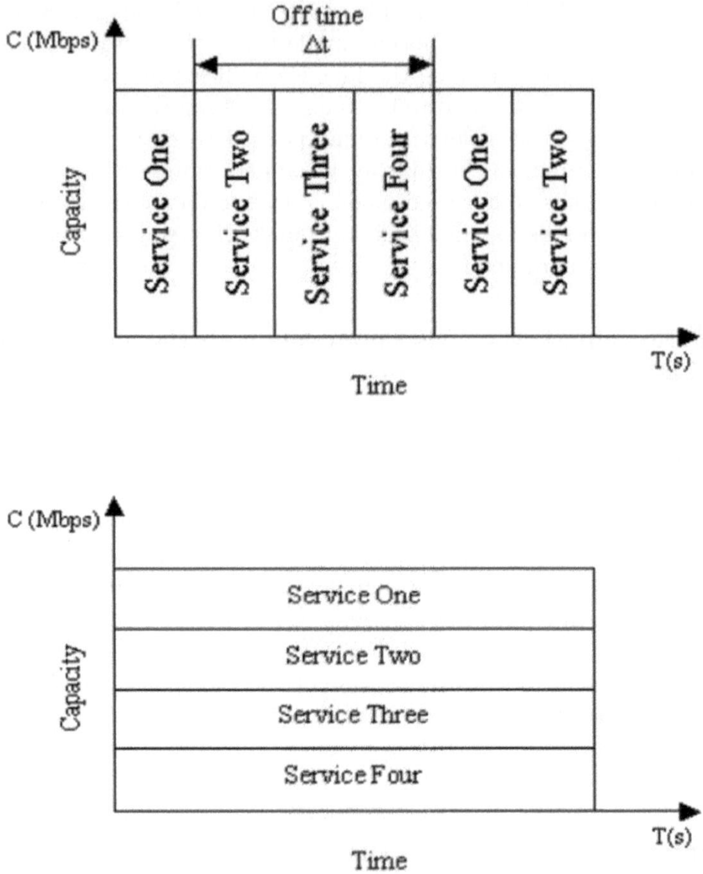

Fig. 10.3. The Time Slicing Structure and the Continuous Transmission Structure

currently active subscribers [46]. The throughput based loading status is chosen for handover from the UMTS network part in order to be coherent with the handover criteria of the DVB-H network part. The UMTS downlink load factor is estimated based on the sum of the bit rates of all the currently active connections in a cell divided by the specified maximum throughput for the same cell:

$$\eta_{DL} = \frac{\sum_{k=1}^{N} R_k}{R_{max}} \tag{10.1}$$

Where R_k is the bit rate of connection k and N is the total number of connections in the dedicated channel DCH in one cell. R_{max} is the maximum allowed throughput of the DCH in the cell. Since the UMTS soft handover can only be supported in the dedicated channel DCH [46], the downlink bit rates from the common channel FACH (Forward Access Channel)and the shared channel DSCH (Downlink Shared Channel)will not be considered.

On the other hand, the down link load factor in DVB-H is expressed by:

$$\theta_{DL} = \frac{\sum_{i=1}^{M} P_i}{P_{max}} \tag{10.2}$$

Where Pi is the bit rate of data channel i, M is the total number of data channels in the DVB-H cell and P_{max} is the maximum allowed data throughput in the DVB-H cell. Before considering the DVB-H side handover criteria, the DVB-H load factor must be kept below the threshold value:

$$\theta_{oldDL} + \triangle\theta \leq \theta_{thresholdDL} \tag{10.3}$$

Where θ_{oldDL} is the DVB-H network load before the user request, $\theta_{thresholdDL}$ is the preset load threshold and $\triangle\theta$ is the load increase after a new data channel is used to accommodate new users.

If the air interface loading is allowed to increase excessively, the quality of service of the existing connections cannot be guaranteed. The admission control principle can assist handover decision-making in the converged network. The admission control algorithm is executed when a new connection is requested or an existing connection is modified.

From the UMTS side, when a new subscriber seeks access to the network, the admission control algorithm estimates the network load and based on the new expected load, the subscriber is either admitted or handed over to the DVB-H system with the group of UMTS users that the subscriber belongs to. In this way, the network load is a kind of threshold triggering the handover. In this scheme the UMTS network can maximize the network usage within a set of network quality levels, i.e. levels depending on what kind of service/information the subscriber wants to use.

The throughput based downlink admission control strategy in UMTS is that a new subscriber is admitted only if the total load after admittance stays below the originally assigned threshold value.

$$\eta_{oldDL} + \triangle L \leq \eta_{thresholdDL} \tag{10.4}$$

Where η_{oldDL} is the UMTS network load before the user request estimated using equation (10.1), $\eta_{thresholdDL}$ is the preset load threshold and $\triangle L$ is the load increase that can be calculated as:

$$\triangle L = \frac{1}{1 + \frac{W}{v \cdot E_b/N_O \cdot R}} \tag{10.5}$$

Equation (10.5) is derived in chapter 9 of [47]. In equation (10.5), W is the chip rate, R is the bit rate of the new user, E_b/N_o is the assumed effective signal to noise ratio of the new connection and v is the assumed activity factor of the new connection representing the burstiness of the offered load. Since only data services are considered in the converged network, v is assumed to be 100%. Therefore, $\triangle L$ for the UMTS network can be expressed as:

$$\triangle L_{umts} = \frac{1}{1 + \frac{W}{E_b/N_0 \cdot R}} \tag{10.6}$$

From the DVB-H side, when a subscriber leaves or enters the DVB-H network, besides the satisfaction of the load status from equation (10.3) the admission control algorithm for DVB-H has to decide whether the number of users receiving the same service drops below a threshold $M_{thresholdDL}$. If there are too few users utilizing DVB-H, the DVB-H network will inform the SMS server to switch these users to UMTS. The algorithm concerning threshold $M_{thresholdDL}$ can be expressed as:

$$M_{oldDL} - \triangle L_{dvbh} \geq M_{thresholdDL} \tag{10.7}$$

Where M_{oldDL} is the number of subscribers requesting the same data service in DVB-H before the subscriber leaves the DVB-H network. $M_{thresholdDL}$ is the preset subscriber number threshold for one channel and $\triangle L_{dvbh}$ is the decrease of the number of subscribers for the specific data service. The number of subscribers in the DVB-H network can be obtained from UMTS network. And $M_{thresholdDL}$ will be obtained later in this chapter using a stochastic tree model.

Equations (10.3), (10.4) and (10.7) denote the intersystem handover criteria equations.

When the throughput based network load in UMTS increases above the threshold $\eta_{thresholdDL}$, the UMTS network will instruct the users requesting the same service to handover to DVB-H. When the number of users receiving the same service in DVB-H drops below a threshold $M_{thresholdDL}$, the UMTS network will instruct these users to switch from DVB-H to UMTS.

One point needing attention is that even when the UMTS network is not heavily loaded it can be still profitable to handover to DVB-H because using UMTS to deliver the same content to mass users is more expensive than using DVB-H and the network cost is not solely depending on whether the UMTS network is heavily loaded or not. The threshold for handover thus should be based on both economic and technical analysis in the network planning stage. The broadcast company and the telecommunications company have to jointly make a decision on the criteria for handover. Details of the tradeoff between the economic and technical issue are discussed later in this chapter.

10.4.4 Handover Execution between UMTS and DVB-H

The intersystem soft handover is a network initiated and mobile assisted handover scheme. The handover algorithm execution between DVB-H and UMTS is performed through the cooperation of the two networks. In the converged network, the Service Management Server (SMS) is the key component for intersystem soft handover management between UMTS and DVB-H. It makes the final decision for intersystem soft handover. The handover procedures are described as follows:

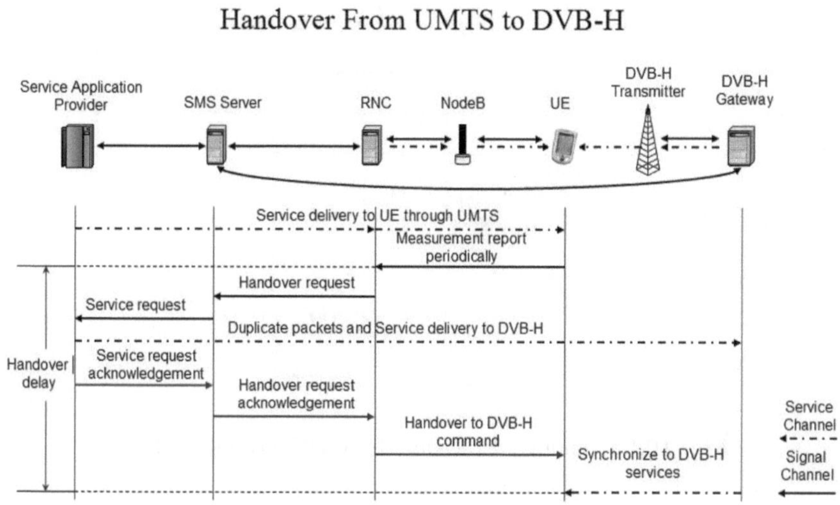

Fig. 10.4. Handover Procedure from UMTS and DVB-H

The first scenario is the handover from UMTS to DVB-H, as shown in Fig. 10.4. Suppose the UE is receiving service delivery from UMTS. The RNC (Radio Network Controller)within UTRAN (UMTS Terrestrial Radio Access Network)receives a measurement report from the UE through NodeB (UMTS Base Station)periodically. When the RNC finds any existing UE requesting a new service or a new UE requesting to join in the UMTS network, it will make a handover judgment based on the handover criteria and the measurement report from the UE. If the decision is to handover the UE to DVB-H, the RNC will send handover request messages to the SMS server. If the SMS server accepts the handover request from UMTS, it will send service request messages to service application provider. Then the service application provider sends service request acknowledgement messages back to SMS server. Then the SMS server will send the handover request acknowledgement messages back to the RNC. Before the service application provider sends service request acknowledgement messages to SMS server it will duplicate and deliver the requested service to the targeted DVB-H gateway through DVB-H. And the DVB-H gateway will deliver the requested services to the UE. When the RNC receives the handover request acknowledgement messages from the SMS server it sends handover commands to the UE. The UE then release resources from UMTS and performs handover to DVB-H.

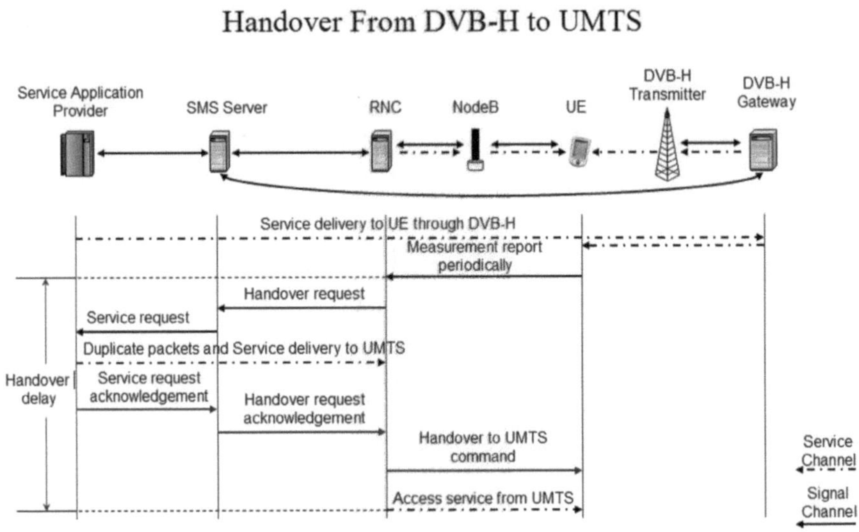

Fig. 10.5. Handover Procedure from DVB-H and UMTS

The second scenario is the handover from DVB-H to UMTS, as shown in Fig. 10.5. In this scenario it is supposed that the UE is receiving file delivery from the DVB-H network. When the number of UEs in the DVB-H cell falls below the preset threshold value, this information in the periodic reports is transmitted through NodeB to the RNC. The RNC will make a handover judgment based on the handover criteria and the measurement reports from each of the UEs. If the decision is to handover the UEs from DVB-H to UMTS, the RNC will send handover request messages to the SMS server. If the SMS server accepts the handover requests from UMTS, it will send service request messages to the service application provider. Then the service application provider will send service request acknowledgement messages back to the SMS server. Then the SMS server will send the handover request acknowledgement messages back to the RNC. Before the service application provider sends service request acknowledgement messages to SMS server it will duplicate and deliver the requested service to the targeted RNC through the UMTS network. And the RNC will unicast the requested services to the targeted UE. When the RNC receives the handover request acknowledgement messages from the SMS server it sends handover commands to the UE. The UE then release resources from DVB-H and performs handover to UMTS.

10.4.5 Handover Performance Evaluation

There are a lot of performance parameters that need to be evaluated for this intersystem soft handover algorithm, such as handover delay, packet loss, handover failure probability, etc. The handover delay is the time duration from when the UE sends the periodic measurement report which contains the handover triggering information to the RNC until the UE get access to the targeted service either from the UMTS or DVB-H network, as shown in Fig. 10.4 and Fig. 10.5 From Fig. 10.4 and Fig. 10.5 it can be seen that the targeted service is delivered towards the UE before the UE hands over to the targeted network. Although such service duplication adds overhead to the network, the handover can be considered seamless and with negligible packet loss. In this section only the handover failure probability of the proposed handover algorithm is evaluated as space does not allow all the parameters to be evaluated in the chapter and handover failure is the most fundamental limitation on the proposed algorithm. The other parameters will be evaluated in subsequent work.

Handover failure happens when the user needs to handover to the targeted network according to the handover criteria but fails to do so, mainly because the targeted network resources are all occupied and cannot provide more channels to accommodate more users. A Markov chain model is used to evaluate the handover failure probability of the proposed handover algorithm here.

A UMTS and DVB-H overlap cell area is taken as an example analysis area. To simplify the analysis, it is assumed that there are a maximum of m data channels in DVB-H and a maximum of n data channels in UMTS in the area. One channel refers to one stream of service delivery. The handover from UMTS to DVB-H and from DVB-H to UMTS are analyzed separately.

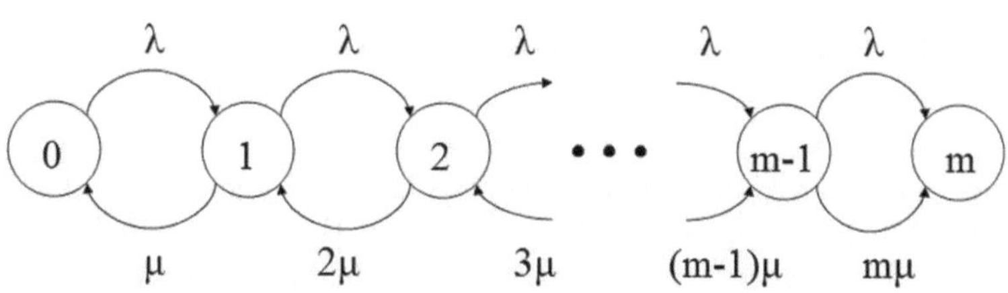

Fig. 10.6. Markov Chain for Handover from UMTS to DVB-H

When handover from UMTS to DVB-H happens, one or several channels of UMTS will be released and only one DVB-H channel will be occupied. The

DVB-H data delivery channels can be modelled by an M/M/m/m queuing process. The state transition diagram is shown in Fig. 10.6, where state j means that there are j busy DVB-H channels in the area. λ_j and μ_j are the birth rate and death rate at state j, respectively.

$$\lambda_j = \lambda, j = 0, 1, 2, \cdots, m-1; \mu_j = j\mu, j = 1, 2, \cdots, m \tag{10.8}$$

When $0 \leq j < m$ a state j will change to $j+1$ if a handover to DVB-H request arrives in the area. Similarly when a handover to UMTS request arrived in the area, the state $j(j > 0)$ will change to$j-1$. According to Erlang's loss formula [18], the handover from UMTS to DVB-H failure probability in the area is

$$P_{failure} = P(m) = \frac{A^m}{m!} / \sum_{j=0}^{m} \frac{A^j}{j!} \tag{10.9}$$

Where $P(m)$ is the steady state probability that the system is at state m and $A = \lambda/\mu$ is the offered traffic intensity. λ is the birth rate and μ is the death rate.

When a group of users in UMTS are requesting an already broadcast service in DVB-H, they will be handed over to DVB-H without occupying any DVB-H new channels. Such users do not have a handover failure problem so they are not considered in the above model.

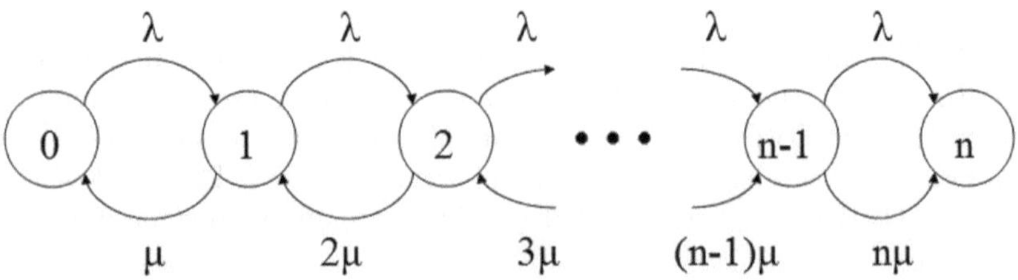

Fig. 10.7. Markov Chain for Handover from DVB-H to UMTS

When handover from DVB-H to UMTS happens, one or several channels of UMTS will be occupied and only one DVB-H channel will be released. The UMTS data delivery channels can be modelled by an M/M/n/n queuing process. The state transition diagram is shown in Fig. 10.7 where state i means that there are i busy UMTS channels in the area. λ_i and μ_i are the birth rate and death rate at state i, respectively.

$$\lambda_i = \lambda, i = 0, 1, 2, \cdots, n-1; \mu_i = i\mu, j = 1, 2, \cdots, n \tag{10.10}$$

Similarly to the handover from UMTS to DVB-H, according to Erlang's loss formula [18] the handover from DVB-H to UMTS failure probability in the area is

$$P'_{failure} = P(n) = \frac{A_1^n}{n!} / \sum_{i=0}^{n} \frac{A_1^i}{i!} \tag{10.11}$$

Where $P(n)$ is the steady state probability that the system is at state n and $A_1 = \lambda/\mu$ is the offered traffic intensity. So the total intersystem soft handover failure probability P is

$$P = aP_{failure} + bP'_{failure} = a\frac{A^m}{m!} / \sum_{j=0}^{m} \frac{A^j}{j!} + b\frac{A_1^n}{n!} / \sum_{i=0}^{n} \frac{A_1^i}{i!} \tag{10.12}$$

Where a is the intersystem soft handover probability of users from UMTS to DVB-H and b is the intersystem soft handover probability of users from DVB-H to UMTS.

Reasonable values [73] are now put into the parameters in equation (10.12) to obtain some numerical results to indicate the handover failure probability of the proposed algorithm. It is assumed that there are $n = 80$ UMTS DCHs, several UMTS cells, and $m = 100$ DVB-H data delivery channels, and one DVB-H cell in the area. The traffic intensities in the area are assumed to be high at $A = 50$, and $A_1 = 50$.

The intersystem soft handover failure probability, also called the handover blocking probability, is usually acceptable when it does not exceed 2% [18]. A diagram illustrating equation (10.12) is given in Fig. 10.8.

From Fig. 10.8 it can easily be seen that the maximum intersystem soft handover failure probability is much less than 2% (the maximum value for the intersystem handover failure probability of this experiment shown in Fig. 10.8 is about 2.2×10^{-5}). With the same methods, even when the traffic intensity grows to $A = A_1 = 69$, as shown in Fig. 10.9, the maximum intersystem handover failure probability is still acceptable at about 2%. It can also be seen that in the $n = 80$, $m = 100$ scenario, the maximum acceptable traffic intensity is $A = 69$, $A_1 = 69$ if $A = A_1$. Using equation (10.12) the intersystem soft handover failure probability in any scenario can be determined providing that the necessary parameters are known.

10.5 Stochastic Tree Model and Analysis

The above sections presented the intersystem soft handover algorithm based purely on the quality of service aspect, which is the technical aspect. However, maintaining the converged network connection, the quality of service levels and reducing the network operation cost when the subscriber requests begin

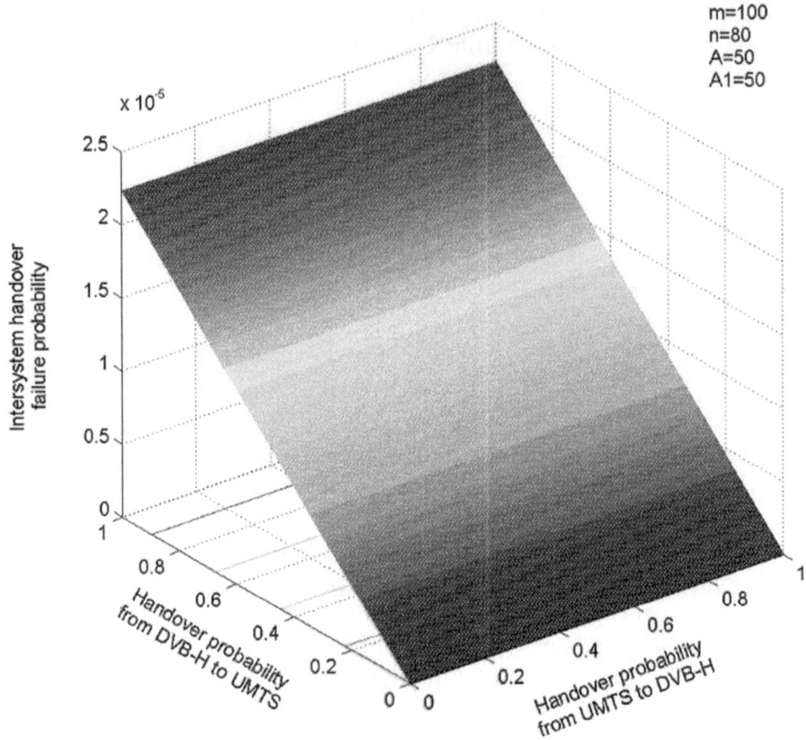

Fig. 10.8. Intersystem Soft Handover Failure Probability ($A = A_1 = 50$)

are all crucial. The costs of network operation can be integrated into the handover algorithm by the network administration. Then to find out whether a user should use DVB-H or UMTS in the scenario of a given service delivery, the tradeoff between the user received quality of service and the network operation cost must be identified. When the optimum tradeoff is found, the threshold value of the handover criteria can be optimized. With this objective in mind, the stochastic tree concept is introduced to analyze the performance of the intersystem soft handover between UMTS and DVB-H in a converged network.

10.5.1 Stochastic Tree instead of Multi-dimensional Markov Chain with Loops

Multi-dimensional Markov chain with loops has long been used in various literatures to mainly calculate the call blocking probability of the state with multiple channels, where the effect of mobility is very well expressed [17] [18]

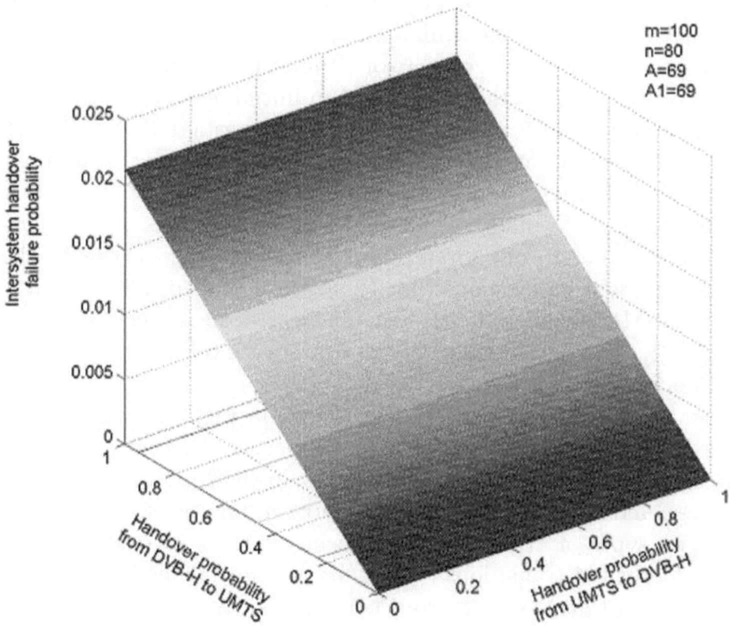

Fig. 10.9. Intersystem Soft Handover Failure Probability ($A = A_1 = 69$)

[103]. However, multi-dimensional Markov chain with loops can not be used to model the cost parameter of the state where the cost is a temporal factor that is increasing in the time domain and is not reversible. Instead, stochastic trees are extensions of decision trees that facilitate the modelling of temporal uncertainties [19] [21]. It is an efficient modelling approach for the decision problems in which the cost and QoS may extend over time. On one hand, this chapter is focusing on the radio resource optimization part of the converged networks which is focusing on the cost and QoS progress in the time domain. On the other hand, in the proposed network structure in the section 10.3 the users will always have available access to both UMTS and DVB-H networks. Thus user mobility is not a concern in the considered scenario. In this way, in order to consider the radio resource optimization decisions on the network side instead of to focus on the user mobilities on the terminal side, stochastic trees are used instead of the multi-dimensional Markov chains with loops.

10.5.2 Stochastic Tree Model for Converged Network

In its simplest and most useful form, a stochastic tree is a transition diagram for a continuous-time Markov chain, unfolded into a tree structure.

Researchers have already used stochastic trees as modelling tools to analyze medical treatment decisions for almost a decade [20]. In this chapter, the stochastic trees are used in the communications field, in making the handover decision, to predict the converged network operation cost and network quality of service that can be expected given both an initial network state and an established handover plan. An advantage of the stochastic tree model is that it allows the recursive computation of the mean quality-adjusted duration, which is the mean duration with quality weights, by "rolling back" the stochastic tree [21] much as one would roll back a decision tree. Before applying the stochastic tree model to our intersystem handover process, the Markovian rollback formula will be developed for this case. Detailed information about the use of the Markovian rollback formula in a medical decision process is provided in [21]. The development of the Markovian rollback formula for the converged network is shown below.

Consider a subtree of a stochastic tree shown in Fig. 10.10 in which an initial state x is occupied until one of several competing transitions with rates λ_y occurs to a state y. The nodes represent states, initialization, UMTS, DVB-H, soft handover, and so on, and the arrows represent transitions between states. There are two types of arrows used in stochastic trees corresponding to the two types of possible transitions. Wavy arrows are labelled with rates and signify transitions that take time to accomplish. Straight arrows are labelled with probabilities and signify immediate transition to one of the states indicated. There are only wavy arrows in Fig. 10.11. Using the wavy and straight arrows representation, the subtree in Fig. 10.11 can be transformed into the subtree in Fig. 10.12 where $p_y = \lambda_y/\lambda$ and $\lambda = \sum_y \lambda_y$. It is apparent that beginning in state x, a mean time $1/\lambda$ is spent in state x, following which transition to a state y occurs with probability p_y.

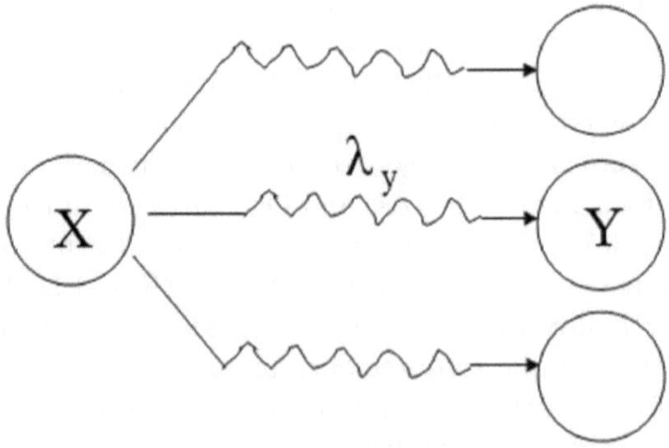

Fig. 10.10. The Stochastic Subtree

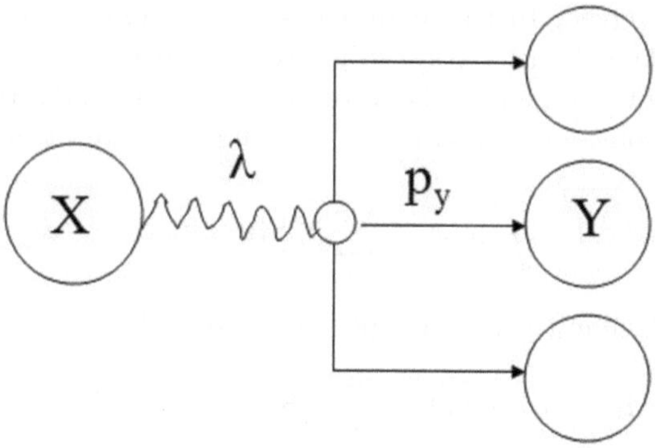

Fig. 10.11. The Stochastic Subtree Transformation

To model the intersystem soft handover in the converged network, a new concept called Quality Service Time (*QAST*) is introduced. This network measurement is defined as being calculated by weighting each time interval T_x spent in a particular network state x by a received service quality factor $v(x)$ that is proportional to the service quality estimation of that network state. This is basically using human subjective perception to provide a value assessment to grade the service quality [36, 37].

$$QAST = \sum_x v(X)T_x \tag{10.13}$$

Where a service quality factor $v(x) = 1$ refers to perfect reception while $v(x) = 0$ refers to no reception, and $v(x)$ takes values between 0 and 1. Suppose the $QAST$ beginning in state y is $S(y)$, then the $QAST$ beginning at x is

$$S(x) = v(x)\frac{1}{\lambda} + \sum_y p_y S(y) = \frac{v(x) + \sum_y \lambda_y S(y)}{\sum_y \lambda_y} \tag{10.14}$$

$$S(x) = \frac{v(x) + \sum_y \lambda_y S(y)}{\lambda} \tag{10.15}$$

Here λ is the rate of departure from state x. $v(x)+\sum_y \lambda_y S(y)$ is the $QAST$ in state x per time unit. This formula can be used to recursively evaluate the $QAST$ in any stochastic tree model. It is called the rollback formula. However, one drawback of the formula is that it is risk neutral. For example, intersystem soft handover between UMTS and DVB-H is not optimum when the duration is too short for the users staying in the handover destination network because a too short duration can make the handover unnecessary and sometimes cause a *Ping Pong* effect. On the other hand, it is advisable to perform intersystem soft handover between UMTS and DVB-H when the user pays a high price to get the optimum service quality or pays the lowest price in spite of the poor service quality received. In this case a risk factor $r(x)$ is introduced as the quality change when the user stays in network state x.

For a Markov process, λ and $r(x)$ represent exponential distributions. Then the $QAST$ after considering risk factor $r(x)$ in state x per time unit is

$$R_t(x_r) = \int_0^t [v(x) + \sum_y \lambda_y S(y)] r(x) e^{-r(x)s} ds \tag{10.16}$$

Where t is the time the system spent in state x. From equation (10.15)

$$R_t(x_r) = [v(x) + \sum_y \lambda_y S(y)][1 - e^{-r(x)t}] \tag{10.17}$$

Then the $QAST$ after considering risk factor $r(x)$ in state x is

$$R(x_r) = \frac{R_t(x_r)}{r(x)} = \frac{v(x) + \sum_y \lambda_y S(y)}{r(x)}[1 - e^{-r(x)t}] \tag{10.18}$$

In this case, the total $QAST$ after considering risk factor $r(x)$ is

$$S(x_r) = \int_0^\infty R(x_r)\lambda e^{-\lambda t} dt = \frac{v(x) + \sum_y \lambda_y S(y)}{r(x) + \lambda} \tag{10.19}$$

When the user performs handover between UMTS and DVB-H, not only will risk happen, sometimes $QAST$ can be increased. For example, when the user switches from UMTS to DVB-H, they benefit from the high bit rates advantages of DVB-H. On the other hand, when the user switches from DVB-H to UMTS, they may benefit from the gained interaction characteristics of the service. So it is necessary to define a quality bonus variable $B(x)$ for the system.

After considering the quality bonus variable $B(x)$, the total $QAST$ becomes

$$S(x_r) = \frac{v(x) + \sum_y \lambda_y [B(y) + S(y)]}{r(x) + \lambda} \tag{10.20}$$

Since risk and bonus factors are always considered in the converged network systems simply use $S(x)$ instead of $S(x_r)$ for the measurement $QAST$. Then

$$S(x) = \frac{v(x) + \sum_y \lambda_y [B(y) + S(y)]}{r(x) + \lambda} \tag{10.21}$$

As $\lambda = \sum_y \lambda_y$, to make calculations easier and practical, equation (10.20) is written as

$$S(x) = \frac{v(x) + \sum_y \lambda_y [B(y) + S(y)]}{r(x) + \sum_y \lambda_y} \tag{10.22}$$

$QAST$ is the measurement of the quality of service of the system. Another measurement $COST$ needs to be defined to evaluate the converged network operation cost. Similarly to the definition of $QAST$, the $COST$ is defined as being calculated by weighting each time interval T_x spent in a particular network state x by a network operation cost factor $c(x)$ proportional to the network operation cost estimated for that network state. So $COST$ is defined as:

$$COST = \sum_x c(x)T_x \tag{10.23}$$

Where a service quality factor $c(x) = 1$ refers to the maximum cost while $c(x) = 0$ refers to no cost, and $c(x)$ takes values between 0 and 1. Similarly to equation (10.21), the total $COST$ for the converged network after considering risk factors and bonus factors can be calculated as:

$$C(x) = \frac{c(x) + \sum_y \lambda_y [B(y) + C(y)]}{r(x) + \sum_y \lambda_y} \tag{10.24}$$

Where $C(x)$ is the total $COST$; $B(y)$ and $r(x)$ are the bonus and risk factor respectively.

10.5.3 Stochastic Tree Model for Intersystem Soft Handover

To use stochastic trees to model the intersystem soft handover for the converged network the following assumptions are made:

1. The network state in the converged network is Markovian property. This means that the next network state is only dependent on the present network state and has no dependency on the previous network states.
2. The quality service factor $v(x)$ and network operation cost factor $c(x)$ are not depending on the number of users in the two networks, they are the statistical values that can be provided by the network operators.
3. The quality service $QAST$ is larger in a UMTS state than in a DVB-H state when a single user receives the same bit rates service but the service has an interactive nature.
4. The quality service $QAST$ is smaller in a UMTS state than in a DVB-H state when the number of users receiving the same service exceeds the preset threshold because of the congestion risk due to the unicast nature of UMTS.
5. The network operation cost $COST$ is higher for DVB-H than for UMTS when the same service is delivered to a single user because the bandwidth is dedicated to one user in this case.

6. The network operation cost $COST$ is higher for UMTS than for DVB-H when the same service is delivered to a certain number of users in the case that the user number exceeds the preset threshold because of the unicast nature of UMTS.
7. The soft handover state is a special state in the converged network. It by default means the intersystem soft handover state between UMTS and DVB-H. The time the UE stays in the soft handover state is not negligible. The states after the soft handover state could be either staying in the original state or entering into the handover destination state. The network operation cost is higher in the soft handover state than in the other states. The service quality is lower in the soft handover state than in the other states.
8. User initiated intersystem soft handover is not permitted so the focus is on the optimum trade-off between network operation cost and quality of service from the network side.

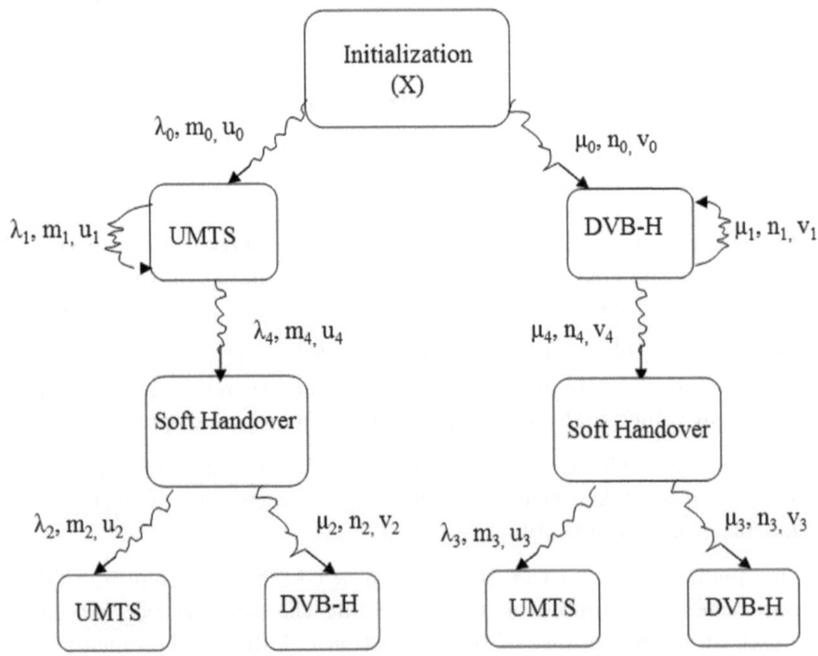

Fig. 10.12. Intersystem Soft Handover Stochastic Tree

Under these assumptions, a stochastic tree model is set up for the inter-system soft handover and is shown in Fig. 10.12. In Fig. 10.12, there are four different states. The transition probabilities are denoted by λ_i and μ_i (i=0, 1, 2, 3, 4). For each network state the $COST$ rates are denoted by m_i and n_i (i=0, 1, 2, 3, 4) and the $QAST$ rates are denoted by u_i and v_i (i=0, 1, 2, 3, 4). Using equations (10.21) and (10.23) the expected total $QAST$ and the expected total $COST$ can be expressed as

$$E[Q(X)] = \frac{v_x + \sum_y \lambda_y (B(y) + E[Q(y)])}{\beta_x + \sum_y \lambda_y} \tag{10.25}$$

$$E[C(X)] = \frac{c_x + \sum_y \lambda_y (B(y) + E[C(y)])}{\alpha_x + \sum_y \lambda_y} \tag{10.26}$$

Where X is one of the states in the stochastic tree, c_x is the cost weight for state X, v_x is the quality weight for state X, a_x is the state-specific cost discount rate at state X, β_x is the quality decrease attitude in state X. $B(y)$ is either $COST$ rate (m_i, n_i) or $QAST$ rate (u_i, v_i) depending on the network state. $E[Q(y)]$ and $E[C(y)]$ are the mean value of $COST$ and $QAST$ in the state after state X respectively.

10.5.4 Simulation and results

In this section, a numerical simulation and its analysis to find the optimum trade-off between the converged network operation cost $COST$ and the quality of service time $QAST$ are presented. The simulation is done based on the stochastic tree structure shown in Fig. 10.12. From Fig. 10.12, for simplicity it is reasonable to suppose that the following equations are true: $m_0 = m_1 = m_2 = m_3; n_0 = n_1 = n_2 = n_3; m_4 = n_4 = m_0 + n_0; u_0 = u_1 = u_2 = u_3; v_0 = v_1 = v_2 = v_3; u_4 = v_4 \leq min(u_0, v_0)$.

Using equations (10.24) and (10.25), the expected total $QAST$and the expected total $COST$ for both UMTS and DVB-H can be calculated. Therefore, the trade-off between $QAST$ and $COST$ can be obtained. Thus, the network side handover criteria can be set up based on the trade-off between the converged network operation cost and the quality service time.

The network scenario in the simulation is an HTTP file delivery to a single user in the converged network. The HTTP service can be received through both the DVB-H and UMTS networks individually. Some interaction features are available if the service is delivered through UMTS. First the variables and their values used in the numerical simulation are given. These parameters are shown in Table 10.1 below.

For simplicity it is also supposed that the risk factors $\alpha_x = \beta_x = 0$; The mean values for $COST$ and $QAST$ in the last UMTS state and DVB-H state of the stochastic tree are $E[Q(UMTS)] = u_2 = u_3$, $E[Q(DVBH)] = v_2 = v_3$, $E[C(UMTS)] = m_2 = m_3$,$E[C(DVB-H)] = n_2 = n_3$. For simplicity

Table 10.1. Probabilities Used in the Trade-off Analysis

Description	Variable Name	Value
From Initialization to UMTS network probability distribution	λ_0	0.5
	m_0	0.2
	u_0	0.6
From Initialization to DVB-H network probability distribution	μ_0	0.5
	n_0	0.8
	v_0	0.5
Soft handover probability distribution from UMTS to soft handover state	λ_4	0.5
	m_4	1
	u_4	0.4
Soft handover probability distribution from DVB-H to soft handover state	μ_4	0.5
	n_4	1
	v_4	0.4
Probability distribution from UMTS originated soft handover state to UMTS	λ_2	0.5
	m_2	0.2
	u_2	0.6
Probability distribution from UMTS originated soft handover state to DVB-H	λ_2	0.5
	n_2	0.8
	v_2	0.5
Probability distribution from DVB-H originated soft handover state to UMTS	λ_3	0.5
	m_3	0.2
	u_3	0.6
Probability distribution from DVB-H originated soft handover state to DVB-H	μ_3	0.5
	n_3	0.8
	v_3	0.5
Probability distribution staying in UMTS network state	λ_1	0.5
	m_1	0.2
	u_1	0.6
Probability distribution staying in DVB-H network state	μ_1	0.5
	n_1	0.8
	v_1	0.5

reasons, the parameter values used in this simulation are very simple and are just for illustration, so they may not reflect the real field test data. However, they are reasonable and can provide a general idea of how the algorithm works. Practical parameter values can be used in the model after real field tests in the network planning and optimization stage. From equations (10.24) and (10.25), the expected total $COST$ and the expected total $QAST$ for UMTS and DVB-H in the case of one user can be calculated along the two separate trunks of the stochastic tree: $E[C(X)]_{umts} = 5.4; E[C(X)]_{dvbh} = 3.6; E[Q(X)]_{umts} = 3.6; E[Q(X)]_{dvbh} = 3.3$.

For N users of the same service, suppose N_1 users are using UMTS and N_2 users are using DVB-H the total COST and QAST for the converged network are: $C_{total} = 5.4N_1 + 3.6$; $Q_{total} = 3.6N_1 + 3.3N_2 = 0.3N_1 + 3.3N$. In order to find out the network performance trade-off point, suppose $C_{total} = Q_{total}$, then optimum number N_1 can be expressed as:

$$N_1 = (11N - 12)/17 \tag{10.27}$$

The trade-off equation (10.26) is drawn in Fig. 10.13. Suppose that 500 users are served by the converged network with the same network parameters in the above simulation in the area, the optimum number of users served by UMTS file delivery will be 323.

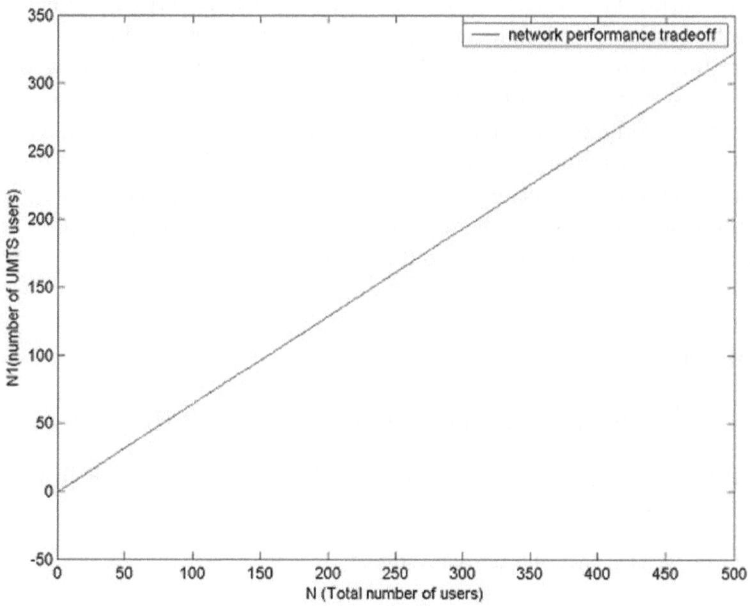

Fig. 10.13. Number of UMTS Users vs. Number of Total Users

Using the presented stochastic tree model, the intersystem soft handover decision in the converged network can be made from both the technical and economic point of view. The model given above is not supposed to be complete, however, it gives a general idea of how to find the optimum trade-off to perform intersystem handover in converged networks where the technical issue is only one part of the problem and the cost elements have to be considered. For different network planning and optimization strategies different stochastic tree models can be set up.

10.6 Conclusions

A converged network on a common IP core is a trend for the next generation communication systems. A DVB-H network and a UMTS network have different radio resource allocation processes and different handover mechanisms. In a converged DVB-H and UMTS network, cooperation between DVB-H and UMTS must happen to allocate radio resources and perform intersystem handover. This cooperation not only includes technical aspects that maximize the network performance and provide the best quality of service (QoS)to users but also includes economic aspects that maximize the operating profit for both telecommunication and broadcast network operators.

This chapter has described, for a converged DVB-H and UMTS network, a novel intersystem soft handover algorithm. To consider the intersystem handover criteria from both technical and economic aspects, two new network measurement parameters $QAST$ and $COST$ were defined. Stochastic tree concepts were introduced and a stochastic tree model for the converged network was presented. A numerical simulation was used to illustrate the intersystem soft handover process between DVB-H and UMTS. From the simulation results, the trade-off between the network operation cost and the quality service time can be found for the converged network.

Problems

10.1. What is the main function of the converged DVB-H/UMTS networks?

10.2. What is the intersystem handover in the converged DVB-H/UMTS networks?

10.3. Is DVB-H a broadcast technology?

10.4. How can DVB-H and UMTS complement each other?

10.5. What is the cell structure of the converged DVB-H/UMTS networks?

10.6. What features does the terminal need to have in order to perform intersystem handover in the converged DVB-H/UMTS networks?

10.7. How does the terminal perform intersystem soft handover measurement in the converged DVB-H/UMTS networks?

10.8. What is a stochastic tree?

10.9. Why is the stochastic tree used in the converged DVB-H/UMTS networks?

11

Handover Algorithm for Hybrid Broadcast Networks

11.1 Introduction

DVB-H is a mobile broadcast standard. There are other different mobile broadcast standards being developed and deployed around the world.

1. Terrestrial - Digital Multimedia Broadcasting (T-DMB)
   - T-DMB is developed in Korea and is based on the DAB standard. DAB already has time slicing features and it was planned to be used in a mobile environment. However, basic DAB's error protection is insufficient for video services and T-DMB adapts DAB by adding additional error protection, the Reed-Solomon error correction coding. In addition, T-DMB has incorporated AAC+ audio codec.
2. Digital Multimedia Broadcasting - Terrestrial (DMB-T)
   - The core technique of DMB-T is Time Domain Synchronous Orthogonal Frequency Division- Multiplex (TDS-OFDM) modulation, at the same time it also implements better FEC (that is LDPC code) and adopts hierarchical modulation structure for different services. At the physical layer, it provides a very flexible channel coding rate with the combination of different modulation constellations to achieve a wide range of bit rates (5 - 33 Mbit/s). DMB-T carries the MPEG-2 Transport Streams containing any combination of video, audio and data. DMB-T chooses the PN sequence as the guard interval of the OFDM symbol to achieve much quicker synchronization (time domain processing). This is very important for the packet switching at a high transmission rate. A similar idea can be found in the WLAN system. DMB-T allows IP multicast and unicast on top of MPEG-2, which lays the foundation of the interactive services. Especially in the IPv6 system, every digital device connected to this network can be assigned an IP address, DMB-T devices can therefore use the IP network for multicasting and unicasting. DMB-T can also work for very high-speed (more than 130km/h) mobile reception with less than 1e-10 BER [14].

3. MediaFlo
   - MediaFlo, also called FLO, is a mobile broadcast standard developed by Qualcomm and is expected to be widely deployed within the USA. It is claimed that FLO offers better performance for mobility and spectral efficiency with minimal power consumption. FLO also utilizes Time Division Multiplexing (TDM) to reduce power consumption. The other features of FLO includes: FLO supports the coexistence of local and wide-area coverage within a single Radio Frequency (RF) channel; layered modulation; incorporation of a turbo inner code and a Reed Solomon (RS) outer code [165].
4. Integrated Services Digital Broadcasting - Terrestrial (ISDB-T)
   - ISDB-T is developed in Japan and is designed to provide various services, including HDTV, multi-channel SDTV, data services, etc. for portable and mobile reception. ISDB-T is characterized by: flexible use of modulation schemes or hierarchical transmission (digital contents can be simultaneously transmitted with the appropriate modulation schemes and appropriate bit-rates for each type of content integrated in the ISDB stream); use of a control signal that informs the receiver of the multiplexing and modulation configuration; partial reception (some of the services can be received by a lightweight, inexpensive narrow-band receiver) [162].
5. Digital Video Broadcasting - Satellite and Terrestrial (DVB-SH)
   - DVB-SH is derived from DVB-T, DVB-H and DVB-S2. It is a transmission system for hybrid satellite and terrestrial digital video broadcasting to mobile terminals. DVB-SH transmission system consists of two parts: a Satellite Component (SC) and a Complementary Ground Component (CGC). The SC ensures geographical global coverage while the CGC provides cellular type coverage. DVB-SH has two different kinds of transmission modes: OFDM mode which is based on DVB-T and TDM mode which is based on DVB-S2. New features incorporated into DVB-SH includes FEC encoding using 3GPP2 turbo code. In the modulation modes, FFT length 1K, 2K, 4K and 8K are specified [89] [90].
6. China Multimedia Mobile Broadcasting (CMMB)
   - CMMB is similar to DVB-SH, but developed and specified in China. CMMB is based on the Satellite and terrestrial interactive multiservice infrastructure (STiMi) which is also developed by China. STiMi works in the frequency range 30MHz - 3000 MHz and supports 2MHz and 8MHz bandwidth. It utilizes Low Density Parity Check (LDPC) channel coding and OFDM modulation. Modulation modes BPSK, QPSK and 16QAM are supported [141].

The rest of the chapter is organized as follows:

Section 11.2 provides an overview of the hybrid broadcast network structure and of the vertical handover issue within the hybrid broadcast networks. Section 11.3 presents the vertical handover algorithm between the different broadcast networks within the hybrid broadcast networks, including the handover measurement, handover criteria, handover execution and evaluation. In Section 11.4, some analysis are presented. Section 11.5 concludes the chapter.

11.2 Hybrid Broadcast Network Overview

Different mobile broadcast standards are being developed and deployed around the world. The future scenario will be different standards coexisting. It is very possible that the users move from one country or area to another and still hope to get the service using the same devices. For this reason, it is necessary to make vertical handover between the different broadcast networks. The hybrid broadcast network is shown in Fig. 11.1.

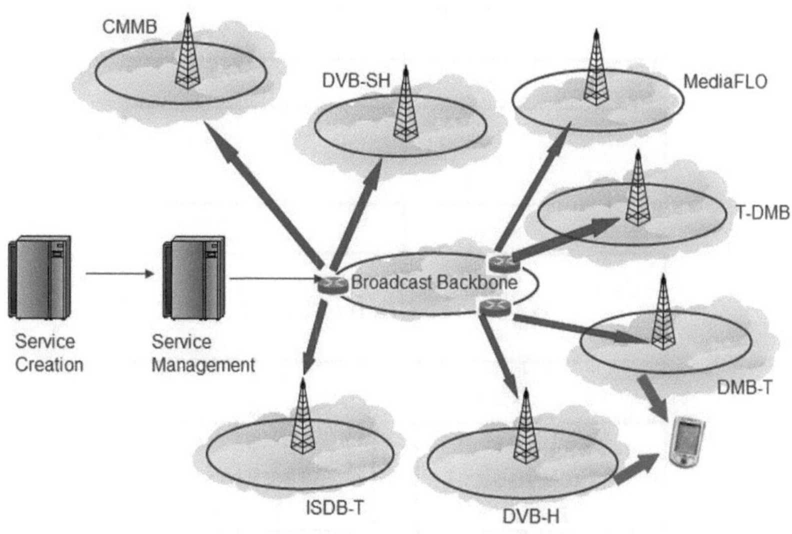

Fig. 11.1. Hybrid Broadcast Network Overview

Fig. 11.1 shows an abstract hybrid broadcast network structure where the terminal has to perform vertical handover between different broadcast network covered area. In this chapter the vertical handover between DVB-H and DMB-T is used as an example to illustrate the handover issue in the hybrid broadcast networks.

11.3 Vertical Handover in the Hybrid Broadcast Networks

The algorithm of the vertical handover in the hybrid broadcast networks is to design a handover control layer between the IP layer and the lower Mac/Physical layer. As the upper layers above the handover control layer are almost the same for different broadcast standards. The main task will be how to detect and convert the signals from the lower layers of the different broadcast networks and to convert them into the IP packets of the common upper layers. The algorithm structure is shown in Fig. 11.2.

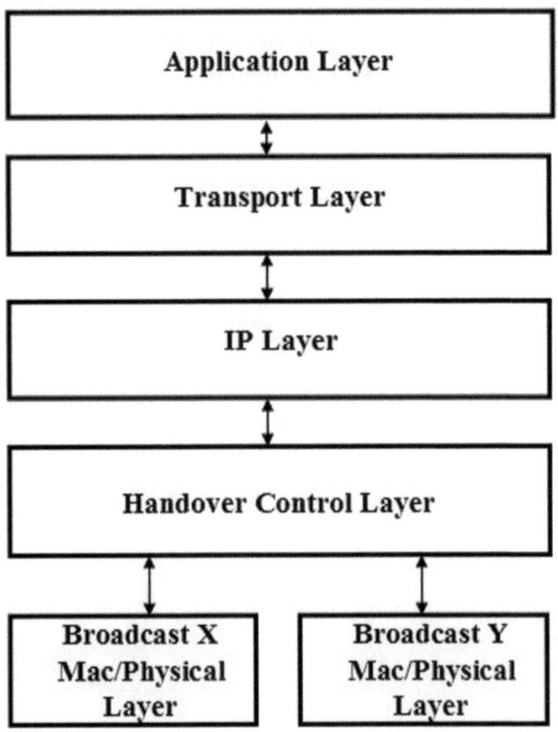

Fig. 11.2. Algorithm Structure for the Handover in the Hybrid Broadcast Networks

Fig. 11.2 shows the location of the handover control layer for the vertical handover between broadcast standard X and broadcast standard Y. In this chapter the vertical handover between DVB-H and DMB-T is used as an example to illustrate the handover issue in the hybrid broadcast networks, which is shown in the following subsection.

11.3.1 Handover between DVB-H and DMB-T

The illustration of the handover between DVB-H and DMB-T is set in the following scenario:

In the hybrid DVB-H and DMB-T networks, the same area is covered by both DVB-H and DMB-T or one area is covered by either DVB-H or DMB-T. The hybrid standard terminal is capable of receiving the services from both of the two networks. When the terminal moves out of one broadcast standard area or the terminal lost signals from one broadcast network, it has to switch to the other broadcast standard network to continue the service reception. Here the services on the two different broadcast networks may not be the same. The terminal in the hybrid broadcast networks is usually a device with uplink communication capabilities.

The process of handover from DVB-H to DMB-T is illustrated in Fig. 11.3.

In Fig. 11.3, each of the broadcast networks (DVB-H and DMB-T) consists of three functional components: Subscription Management, Handover Management and Service Delivery. The Subscription Management component is in charge of terminal registration, authorization and billing. The Handover Management component is in charge of handover request, handover negotiation and handover communication between the different functional components. The Service Delivery component is in charge of service delivery from the Service Application component to the terminals. As shown in Fig. 11.3, the Service Application component is delivering services to the Service Delivery component in both DVB-H and DMB-T networks. When the terminal moves out of the DVB-H coverage area, it detects the existence of the DMB-T signals. The terminal send the service registration information through the uplink (which is not shown in Fig. 11.3) to the Subscription Management component of the DMB-T networks. After the Subscription Management component of the DMB-T networks receives the request, it send service authorization request to the Subscription Management component of the DVB-H networks. Then the Subscription Management component of the DVB-H networks sends the service authorization response if the terminal is allowed to switch to the DMB-T networks. After that the Service Delivery component of the DMB-T networks begins to delivery the service to the terminal. It is easily seen that the communication and cooperation between the functional components of the two broadcast network (DVB-H and DMB-T) are very important.

Similarly the process of handover from DMB-T to DVB-H is illustrated in Fig. 11.4.

The detail handover mechanisms is shown in Fig. 11.5.

As shown in Fig. 11.5, the signalling information PSI/SI and TPS of DVB-H and DMB-T signaling of DMB-T are used by the terminal to decode the signals. The signal quality (RSSI or SNR) are used as handover criteria to evaluate the signal quality of the different networks. After signal switch and resynchronization the terminal is operating in the target networks.

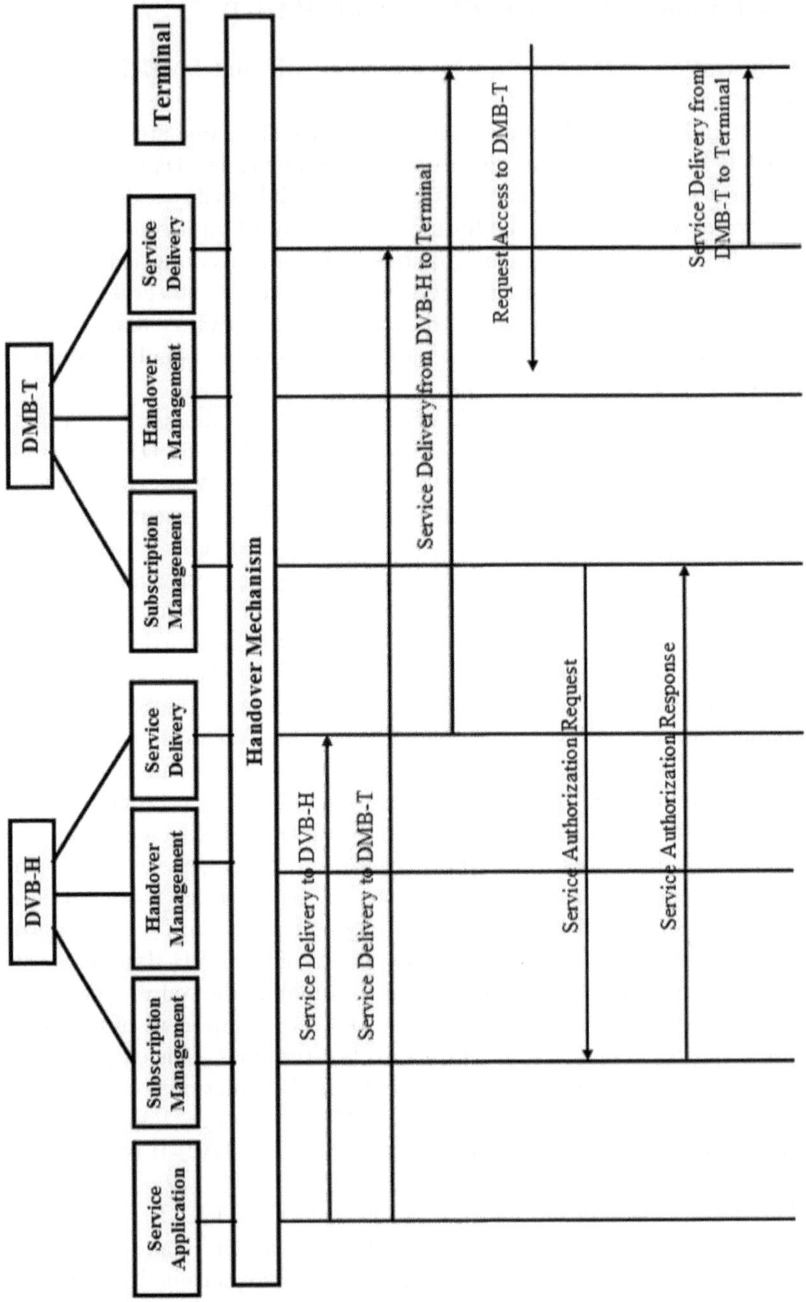

Fig. 11.3. Handover from DVB-H to DMB-T

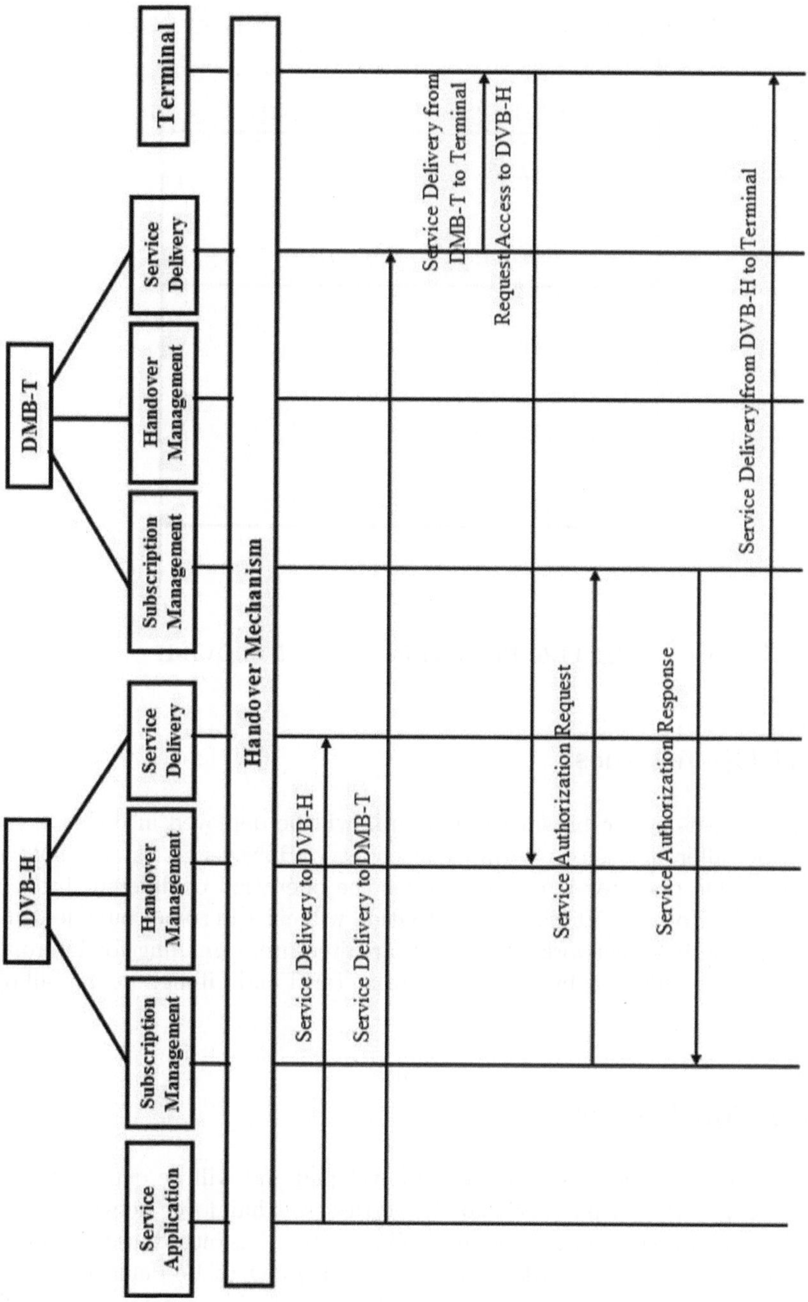

Fig. 11.4. Handover from DMB-T to DVB-H

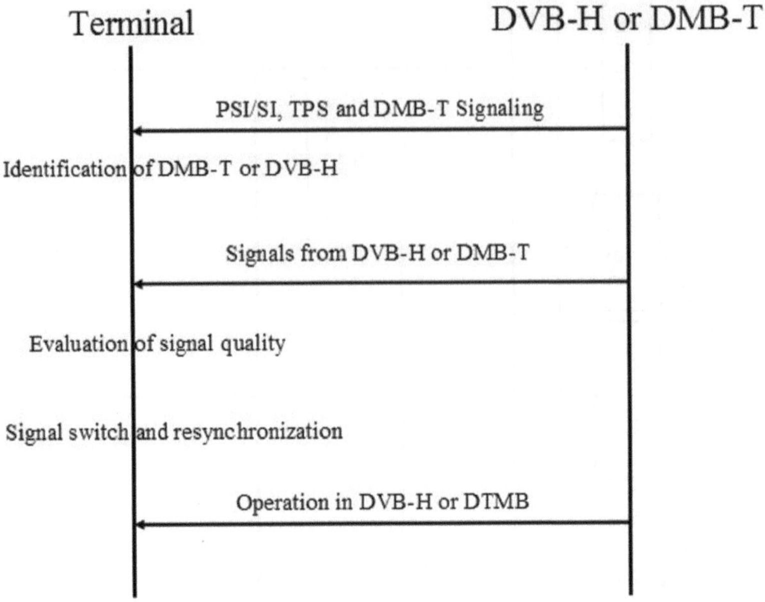

Fig. 11.5. Handover from DMB-T to DVB-H

11.4 Open Issues

The different mobile broadcast standards will be deployed in different countries or different regions within one country. However, it is still not clear whether one operator is responsible for the operation of different broadcast standards. However, at least such situation will exist in some countries. In the hybrid broadcast networks, the detail architecture, signalling for interoperation between different broadcast networks, tariff and billing, etc. are all open issues.

11.5 Conclusions

As different broadcast standards are developed and will be deployed around the world, it is very possible that the vertical soft handover between different broadcast standards will be required. This chapter has introduced the different mobile broadcast standards that are current going to be deployed or being deployed around the world. It is necessary to design a common signalling layer (handover control layer) for the different broadcast standards. However, there are still many open issues regarding to the vertical handover within the hybrid broadcast networks before a practical algorithm is implemented.

Problems

11.1. What is hybrid broadcast network?

11.2. What are the currently main available mobile broadcast standards around the world?

11.3. What is the main idea to design a vertical handover in the hybrid broadcast networks?

12

Conclusions and Future Work

12.1 Conclusions

The research work presented throughout this book are the research results obtained by the author in the past several years. Although there are already some research accomplishments in the DVB-H handover domain, it is still not ready for the technology to be applied in the commercial roll-out due to limited handover field trials and unclear business issues. Therefore, the research results presented in this book are valuable for the commercial implementation of DVB-H and for the future research directions of DVB-H. Each chapter of the book makes one independent reading material for the readers.

Chapter 1 of the book described the general background of the telecommunications and broadcast world, the basic technical features of DVB-H and the book structure. Chapter 2 of the book presented the motivations and approaches of the handover research for DVB-H. In this chapter the readers can understand why and how should we conduct the handover research in DVB-H. The readers will also be able to know how can we design a better handover algorithm for DVB-H. The handover algorithms designed in the rest of the book are mainly based on the methods and guidelines regarding to how to design a handover algorithm in Chapter 2. In chapter 3, a complete survey of handover research in DVB-H was presented. The advantages and disadvantages of different handover algorithms were given. The projects IST INSTINCT and IST MING-T were briefly described as examples of the DVB-H handover related projects. Chapter 4 is focusing on the DVB-H signalling information which includes PSI/SI tables, TPS bits, ESG and EPG. Electronic Service Guide (ESG) is no doubt one of the most important technologies in DVB-H. It is important to and connects the service providers, service operators, terminal manufacturers and users. It differentiates DVB IPDC and OMA BCAST standards, so it has to be considered by the service operators for choosing the appropriate standard. In chapter 6, a preliminary investigation of the different handover algorithms in the dedicated DVB-H networks was presented. A comparison of the different handover algorithms was done and a hybrid handover decision-making algorithm for the dedicated DVB-H networks was illustrated. While chapter 7, 8, 10, 11 are targeting different handover algorithms in different networks (chapter 7 and 8 are for dedicated DVB-H networks; chapter 10 is for converged DVB-H/UMTS networks; chapter 11 is for hybrid broadcast networks), chapter 9 is devoted to the soft handover probability calculation of the RA_handover algorithm.

Handover of the dedicated DVB-H networks will be the first handover algorithm that are going to be implemented. As a short summary for the handover in the dedicated DVB-H networks, Table 12.1 shows the different handover algorithms of the dedicated DVB-H networks presented in this book. Table 12.1 also shows the different algorithms' targeted handover stages and the problems that these algorithms are focusing on. For comparison, the different handover algorithms proposed before by other people are also illustrated in Table 12.1. It takes for granted that Table 12.1 may not be exhaustive regarding to the handover algorithms as it only relates to the handover algorithms that are reported at the time of writing this book.

As an extension to the handover in the dedicated DVB-H networks the handover in the converged DVB-H/UMTS and in the hybrid broadcast networks were illustrated in chapter 10 and chapter 11. In the converged DVB-H/UMTS networks, a novel optimum radio resource allocation algorithm was presented and evaluated. Converged DVB-H and UMTS network is a typical example of incorporating both the broadcast and telecommunications technologies to provide diverse services and to utilize the radio resources more economically. This kind of optimum radio resource allocation algorithm is called intersystem soft handover in this book. The stochastic trees are used in such intersystem soft handover process. Stochastic tree is an extension of decision tree to facilitate the decision making process and it has been used in medical decision-making process for a decade. After the stochastic trees are introduced to the converged network, the quality of the user received service and the network operating cost can be calculated numerically and balanced when a specific requirement for quality of service and network operating cost is made. Analysis and simulation showed that the proposed radio resource allocation algorithm has the potential to be used in the real network environment. Besides DVB-H, other different mobile broadcasting standards are being developed and deployed. In this situation, it is necessary to consider the vertical handover issue between different broadcast standards. Chapter 11 proposes the basic design for a vertical handover algorithm in such hybrid broadcast networks.

DVB-H is a novel technology and hot topic for research. The author believes that the research work and its results presented throughout this book will definitely provide valuable references for the future research in DVB-H in both academic and industry.

Table 12.1. Summary of Different Handover Algorithms

Algorithm	Main Novelty	Handover stages	Problem targeted	Comments
[104]	First paper on handover in DVB-H	Handover measurement, decision-making & execution	Design a feasible handover algorithm	Without mechanisms to prevent the *Ping Pong* effect and "fake signals"
[111]	Cell Description Table (CDT) based handover algorithm	Handover decision-making	*Ping Pong* effect & power consumption	Trade-off between bandwidth (memory) and power consumption should be considered
[105]	Correlation-based "Fast Scattered Pilot Synchronization"	Handover execution	Power consumption	Validation by further study and field trials are needed
[113]	Power-based "Fast Scattered Pilot Synchronization"	Handover execution	Power consumption	Validation by further study and field trials are needed
Chapter 6	Investigation and proposal of different handover decision-making algorithms	Handover decision-making	*Ping Pong* effect, "fake signals" & power consumption	Validation by further study and field trials are needed
Chapter 7	Post-processing of SNR values based handover algorithm	Handover decision-making	*Ping Pong* effect, "fake signals" & power consumption	Validation by further study and field trials are needed
Chapter 8 & 9	RA_handover	Handover decision-making	*Ping Pong* effect, "fake signals" and power consumption	Validation by further study and field trials are needed

12.2 Current and Future Research Work

Research and field trials in DVB-H is the first step towards the final convergence between broadcast and telecommunications. A myriad of projects on DVB-H are being carried out around the world. There were already two main European research and development projects regarding to DVB-H. INSTINCT was a European project addressing IP-based Networks, Services and Terminals for Converging Systems, namely convergence between Broadcast networks (DVB-T/H) and mobile Telecommunications networks (GPRS/UMTS) [137]. INSTINCT started on 1 January 2004 and ended in December 2005. MING-T [140] was another European project supported by EU IST. The mission of the MING-T project is to contribute to speeding up of the interoperation of the DVB-H standard and the Chinese broadcast standard DMB-T by validating the technology and providing adequate inputs to forums and standardization bodies. MING-T was planned for 2 years from January of 2007 to December of 2008. Both of these two projects were intending to pave the way for fully commercialization of DVB-H. The author expects that a more comprehensive research for the converged networks should be on the way when the DVB-H technology is fully validated and commercialized.

Besides the research projects, the field trials of DVB-H have already taken place or are in operation in many places. [139] gives an overview of the current DVB-H trials and commercial services around the world. Most of these trials and commercial services are dealing with Single Frequency Networks thus no handover within multi-frequency networks is required.

Although there are some other technologies that are competing with DVB-H nowadays like T-DMB (Terrestrial Digital Multimedia Broadcasting) and MediaFLO (Media Forward Link Only) [75, 149], the future market for DVB-H is anticipated to be huge. In a report by the Helsinki University of Technology [150], the research shows that the maximum achievable adoption level for DVB-H in Finland is at a coarse estimate 63% of the population of Finland. And the sales figures of DVB-H devices in the year 2008 are predicted to be in the order of 10s to 100s of millions [147]. In order to meet market demands further validation especially of the handover in multi-frequency networks needs to be carried out.

TV is the last media missing from the mobile phones. Thanks to DVB-H technology, TV has become a reality on handheld devices like mobile phones and Personal Digital Assistants (PDAs) . With mobile TV as the representative application [40, 154, 58], DVB-H has already been and will still be a very hot research area. Some of the possible future research topics regarding to DVB-H are:

1. Cross Layer Design
   - One research direction of cross layer design is to combine different scalable encoding layers in the application layer and different modulation mechanisms in the physical layer. Scalable coding meas that the video coding in the application can be divided into one base layer and one or more enhancement layers. DVB-H allows hierarchical modulation. This means the base encoding layer and the enhancement encoding layers can be modulated using different modulation schemes. In this way, when the terminal is in the remote area far away from the transmitters or the reception conditions are not good, the terminal can receive only the base encoding layers to ensure minimum reception. When the terminal is near to the transmitters or the reception conditions are good, the terminal can receive both base encoding layer and enhancement encoding layers thus ensuring the optimum reception.
     The above mentioned approach is only one example of the cross layer designs in DVB-H. The other approaches of cross layer design in DVB-H,

such as combining different encoding layers in the application layer and the different code rates of the physical layer can also be exploited.

2. DVB-H Indoor Reception
   - DVB-H is based on DVB-T and DVB-T was desgined for roof-top antenna reception. Therefore, the indoor reception of DVB-H is usually not good enough to satisfy the users' requirements. Different techniques should thus be designed to improve the signal quality of DVB-H indoor reception.
   Indoor repeaters are devices which are easy to install, cheap to purchase and can provide the required indoor reception quality. Here the problems will be how to design optimum indoor repeaters and indoor repeater location planning.
   WLAN can also be used to provide the relay function of the DVB-H indoor receptions. Here the question will be how to use the WLAN in the DVB-H indoor reception scenario. Should WLAN be used as repeaters to first receive the DVB-H signals from the transmitter (In this case, DVB over IP is required thus synchronization between the WLAN transmitted DVB-H signal and the Transmitter transmitted DVB-H signal is not required.) or should WLAN be connected to the central DVB-H service provider to fetch the signals thus bypassing the DVB-H transmitters (In this case DVB over IP is not required and the synchronization between the WLAN transmitted DVB-H signal and the transmitter transmitted DVB-H signal is required.)? This question is still not fully researched. When WLAN is used in the DVB-H indoor reception, the terminal must be also specifically designed to be able to receive the signals from both DVB-H and WLAN.
3. Handover Design
   - For handover research on DVB-H, the future directions will be to improve the quality of service of the handover algorithms that are proposed while further reducing the battery power consumption of the terminals in the proposed algorithms. The convergence between broadcast and telecommunications is a trend because in the converged networks the broadcast networks can provide cheaper ways to delivery the same multimedia heavy-duty services to mass users while the telecommunications networks can provide excellent interactivity and comprehensive billing methods for the users and operators. Since the convergence of broadcast and telecommunications will come sooner or later, the vertical handover algorithms between broadcast (DVB-T/H) and telecommunications (GPRS/UMTS) will be a potential hot research area. Finally, further validations of the different handover algorithms in both the laboratory test and the commercial field trials are needed.

4. Mobile Client Design
   - Whether the DVB-H will be successful or not is heavily dependent on the acceptance of the customers to the DVB-H services. The direct interface to the customers is the mobile client which can be located at mobile phones, PDAs or other mobile portable devices. A good mobile client should be easy to be used. It should also be transparent to the users regarding to the technology behind the mobile client. The customers are caring more about of the services they are consuming instead of the technologies behind them. For example, a good mobile client on the mobile phone should integrate both the 3G interface and DVB-H interface so that the users can experience seamless consuming experiences. When the users switch between the different TV programm channels on the mobile phone, the users expect a similar operation experiences just like they are using the remote control device for their home based TV sets. The mobile client design is up to now still an open issue and a good mobile client design will win the customers thus making the DVB-H technology success.

As the DVB-H technology is evolving to the next stage DVB-H2, new research issues are also coming up, which will in turn push the technology to the next stage.

Problems

12.1. Please give some examples of future research topics for DVB-H.

Solutions

Problems of Chapter 1

1.1 DVB-H is the product of DVB's process of analyzing commercial requirements and producing an open standard to meet these requirements. Traditional DVB-T technology is not suitable for the efficient and robust transmission to the battery-powered, handheld devices. DVB-H operates in the IP environment, which also makes it a flexible and future-proof technology.

1.2 DVB-H is a broadcast technology. However, it targets the mobile portable devices (like the mobile phones) which are traditionally belonging to the telecommunications domain. For this reason, both broadcast operators and mobile telecommunications operators are looking at DVB-H as a promising technology. DVB-H makes the broadcast world and the telecommunications world converged, making a new network structure (converged networks) come to life.

1.3 DVB-H is based on DVB-T but is for broadcasting IP data and targeting the handheld devices. DVB-H introduces new features: Time slicing, MPE-FEC, 4K mode, 5MHz bandwidth, the use of frequency spectrum outside of UHF (e.g. the L Band) and some DVB-H enhanced signaling information (e.g. Cell Identifier).

1.4 Handover is very important in the cellular mobile networks where the handover of the signals is necessary when the users move from one cell to another. DVB-H is targeting mobile devices. SFN and MFN are two typical network structures in DVB-H. When the DVB-H receivers move from one cell to another in MFN or move from one SFN to another SFN, handover will happen to keep the service continuation. Time slicing makes handover in DVB-H technically possible.

1.5 Soft handover means there is no service interruption when the terminal performs the handover. Time slicing makes soft handover in DVB-H possible. When the DVB-H receiver performs handover, it will measure the signal strength and make handover decisions in the off burst time. When the next burst comes, the DVB-H receiver will continue the service reception without interruption.

There are two categories of DVB-H terminals. One is DVB-H only device without the telecommunication capability, which is often called unconnected devices. The other category is DVB-H device with telecommunication capability. When the DVB-H terminal relies on the telecommunication link to aid the handover process, this kind of handover is called active handover. When the telecommunication link is not used, it is called passive handover. Handover in the dedicated DVB-H networks and handover for the DVB-H only receivers are passive handover.

1.6 Based on functionality, a basic DVB-H system contains the following functional components: Service Source, ESG and FLUTE Server, Encoder, Proxy, IP Encapsulator, DVB Modulator, Amplifier, Transmission Antenna, Receiver. The Conditional Access is optional for a DVB-H system.

1.7 Free-To-Air or Clear-To-Air are refering to the services that are broadcast without any service protection. Any device can receive these services. Free-To-Air services should not be broadcast within IPsec, SRTP or ISMACryp. In this case, a key stream will normally not be broadcast, but if one is broadcast,it should be ignored by the device.

Free-To-View services are broadcast with service protection, but no billing is made to receive the services. However, reception can be restricted to certain users.

Problems of Chapter 2

2.1 IPDC stands for IP Datacasting. IPDC is a service where digital contents, software applications, programming interfaces and multimedia services are delivered through IP (Internet Protocol) using digital broadcasting.

2.2 IPDC brings benefits to DVB-H:

1. IPDC provides a platform for true convergence of services between DVB-H and cellular telecommunication networks (GPRS/UMTS).
2. IPDC allows the coding to be decoupled from the transport layer,that is, all the different coding techniques can be used above the UDP/IP (User Datagram Protocol/Internet Protocol)layer, thus opening the door to a number of features benefiting handheld mobile terminals including a variety of encoding methods, which only require low power from a decoder (Decoding high bandwidth MPEG-2 encoded streaming video/audio is relatively power consuming).
3. IPDC is relatively insensitive to any buffering or delays within the transmission (unlike MPEG-2), this is because IPDC is utilizing IP protocol

which has long been developed upon the non-quality-assured Internet environment.

4. IPDC is well suited for time-sliced transmission.

2.3 Handover is not always necessary in DVB-H networks. Within a DVB-H Single Frequency Network (SFN), the handover is not required since there are no frequency changes across the DVB-H network. However, in the Multi-Frequency Networks (MFN), the handover is compulsory. When the receiver moves between different SFNs, the handover is also required. Since MFN and multiple SFNs are the typical network structures of DVB-H, the handover becomes a critical issue.

2.4 The physical handover and service handover are differentiated depending on whether the consumed service is changed in the handover process. When the services received by the receiver before and after the handover process are the same, such handover is called physical handover. On the opposite, when the services received by the receiver before and after the handover process are changed, such handover is called service handover. The reason for service handover is usually that the previous received service before handover is not available when the receiver finishes the handover process. In this situation, another alternative service is captured by the receiver.

2.5 Hadover process in DVB-H is normally divided into three stages: handover measurement, handover decision-making based on the handover criteria, and handover execution.

Handover measurement stage is the phase when the receiver measures the signal strength, the SNR or other values of the current cell and the adjacent cells according to the handover criteria. Handover decision-making stage is the phase when the receiver makes the handover decision on whether and when the handover should happen based on the handover measurement obtained from the first stage. Handover execution stage is the last phase of the handover process where the handover to the target signal is carried out.

2.6 Handover in DVB-H is a novel issue. There are many challenges for the handover algorithm in DVB-H. Basically there are four challenges: the *Ping Pong* effect, "fake signals" or tuning failure, excessive power consumption and packet loss.

Ping Pong effect refers to the fact that the handover happens repeated because of the signal fluctuation. The *Ping Pong* effect happens usually in the cell border area.

"fake signals" or tuning failure refers to the fact that the receiver makes handover to the wrong signal which has the same frequency and cell_id as the targeted correct signal.

Excessive power consumption in DVB-H handover refers to the fact that excessive handovers or excessive handover measurements take place which should be avoided.

Packet loss refers to the fact that some packet loss happens during the handover process. DVB-H is using time slice mode for transmission. Even a single packet loss may result in the whole time slice unusable. Therefore, packet loss should be optimally controlled.

2.7 Besides addressing the handover challenges like the *Ping Pong* effect, "fake signals" or tuning failure, excessive power consumption and packet loss, there are other issues need to be considered in order to design an efficient handover algorithm. These issues are:

Predicting the handover moment in the handover decision-making stage;

Reducing handover complexity to ease the handover receiver design and increasing handover compatibility with the available DVB-H standards;

Utilizing additional signalling information (e.g. UMTS link) to aid the handover process;

Avoiding additional equipment cost which is solely for handover (such as a GPS receiver only for handover purposes).

Problems of Chapter 3

3.1 Since the RSSI value can vary due to multipath, interference or other environmental effects it may not give a true indication of the communication performance or the range and mistakenly measuring the RSSI value would result in the Ping Pong effect in handover measurement consuming power unnecessarily. The RSSI value could be measured many off burst times with the RSSI value being measured at least once every off burst time in the worst case. Constant measuring of the adjacent cells signal level without any handover prediction leads to more battery power consumption. In order to overcome these shortcomings a better handover prediction algorithm had to be developed.

3.2 CDT based handover utilizes a CDT (Cell Description Table) to devide the cell coverage area into 256 different signal levels. Thus the receiver knows its location within the cell coverage area. In order to achieve this, the terminal must have a GPS receiver incorporated.

3.3 The Scattered Pilots are training symbols that form a periodic pattern with specific period in time and in frequency within an OFDM frame. The Scattered Pilots are transmitted at a boosted power level to facilitate the synchronization of the OFDM frames.

3.4 Phase Shifting is a technique used to synchronize the signals of adjacent DVB-H cells to ensure packet loss-free handovers. Phase Shifting means there is always a time shift of synchronization between adjacent cells so that there is enough time interval between neighbouring time slices to avoid the possible packet loss caused by time slice overlap when the terminal moves from one cell to another.

3.5 The DVB Project technical reports [69] and [68] proposed a simple handover algorithm for handover in DVB-H based on the handover algorithm in DVB-T. The basic idea is for the DVB-H terminal to use the terrestrial_delivery_system_descriptor, the frequency_list descriptor, the original_network_id and the transport_stream_id together as a pair along with the service_list_descriptor to decide which frequency and transport stream the receiver should switch to in the handover process. In a word, the DVB Project only defines how to implement the handover according to the PSI/SI tables which is actually required for each handover algorithm. Therefore, it leaves much space for the handover algorithm design and development.

Problems of Chapter 4

4.1 The main DVB-H signalling information are PSI/SI tables (data link layer), TPS bits (physical layer), ESG and EPG (application layer).

4.2 SDP is part of ESG and it conveys the information about the transmitted audio/video services like source IP addresses, ports, audio/video encoding formats, etc.

4.3 There are two different ways to transmit a SDP file. One is called Inline SDP. The other way is called Out of Band SDP. Inline refers to that the SDP file is contained in the SessionDescriptionType element. Out of Band refers to that the SDP file is referenced in the SessionDescriptionType element.

Problems of Chapter 5

5.1 Normally the OMA ESG is relying on the underlying Broadcast Distribution System (BDS). It will be adapted to the underlying different BDSs, e.g. DVB-H, MBMS, etc.

5.2 In the DVB IPDC ESG bootstrap stage, the ProviderID is used to distinguish the different ESG providers. The terminal uses the ProviderID to identify a particular ESGEntry within the ESGAccessDescriptor file, which contains information on how to acquire the ESG.

5.3 If DVB-H is used as a BDS, the DVB-IPDC bootstrap ESG is reused. And the ESGProviderDiscovery descriptor and ESGAccessDescriptor are used to allow the discovery of the provider of the OMA BCAST ESG and the access to the OMA BCAST ESG, i.e. the ProviderID is used to distinguish the different ESG providers.

The ESGAccessDescriptor then links to the Service Guide Announcement Channel in the OMA ESG. Independent of the underlying BDS, the different Operators are also distinguished by using id (bcast://operator Y.com or bcast://operator X.com) and BSMFilterCode (OPERATOR_X or OPERATOR_Y) and TransportObjectID (2 or 9) Id, BSMFilterCode and TransportObjectID are within SGDD.

1. Id
   - Unique identifier of the SGDD within one specific SG
2. BSMFilterCode which contains the following attributes and elements:
   - type
   - serviceProviderCode
   - corporateCode
   - serviceProviderName
   - nonSmartCardCode
   - 3GPPNetworkCode
   - 3GPP2NetworkCode

5.4 Yes. The concept in BCAST is also called "announcing service guides within a service guide".

A single BCAST ESG transport supports the marketing messages of several service operators; a separate ESG for each operator is needed in the IPDC alternative. In addition, OMA BCAST ESG can be adapted to support both DVB-IPDC and BCAST terminals.

5.5 Interactive links can be embedded to the BCAST ESG, but they are not currently standardized in the IPDC ESG In addition, Smartcard Profile of BCAST ESG requires an interactive channel to obtain key material. BCAST client can request specific ESG information or the complete ESG through interaction channel according to the OMA ESG standard.

Regarding to file delivery, the interaction channel in BCAST can be used for file repairing.

The repection status report may be sent through the interaction channel.

5.6

1. In OMA BCAST: Setup a Service Guide with a PurchaseChannel fragment identifying a PortalURL pointing to the entry point of a related web-based system.

2. In DVB IPDC: Same as OMA BCAST, but with different Fragment Semantics

5.7

OMA BCAST:

Terminal scans or otherwise detects available Broadcast Distribution Systems (BDS). Then the Terminal attempts to perform Service Guide discovery bootstrap to locate entry point to BCAST Service Guide on all or any of the detected BDSes. Upon successful completion of bootstrap procedure, the Terminal acquires the entry point to BCAST Service Guide over the respective bearer. Consequently, the Terminal acquires SGDDs either by receiving or by retrieving those.

So the process is like:

BDS Bootstrap - Provider ID - SGServerAddress - SGDD - SGDU

If the Interaction channel is used for Service Guide transmission, the entry point should be a URL which indicates the location of the Service Guide, e.g. http://provider.com/serviceguide. Then the process is like: URL - SGDD - SGDU

DVB-IPDC:

ESG bootstrap is started, then different ESGs are accessed.

So the process is like:

ESG Bootstrap - Provider ID - ESG Acquisition

5.8

About the SDP files:

1. OMA BCAST
   - Using Access Fragment to contain or refer to the SDP files.
2. DVB-IPDC
   - Using the Acquisition Fragment to contain or refer to the SDP files.

(BMCO Profile does not have "inclusion of the SDP file in the access fragment" but "referring to an external SDP".)

5.9 In IPDC ESG there are multicast IP addresses and the EPG information are transmitted using the XML format within FLUTE carousel; in OMA ESG it is located in SGDU and transmitted within FLUTE carousel or interaction channel.

5.10 The Bootstrap information for IPDC ESG BootstrapService is:

The destination IP address is fixed to 224.0.23.14 for IPv4 or FF0X:0:0:0:0:0:0:12D for IPv6. Fixed Destination Port: 9214;

If the BDS bootstrap signaling information is not available for OMA BCAST, the following IP address and port number of the FLUTE session should be used:

Fixed Destination Multicast IP Address: 224.0.23.165 for IPv4 or FF0X:0:0:0:0:0:0:132 for IPv6. Fixed Destination Port: 4090.

5.11 In OMA BCAST:

Setup a Service Guide with a PurchaseChannel fragment identifying a PortalURL pointing to the entry point of a related web-based system. The purchase information is delivered within SGDU.

5.12 There are two ways to compress the ESG Fragments: compression using GZIP and compression using BiM. Besides, the ESG Fragments can also be transmitted without compression.

Problems of Chapter 6

6.1 Handover in dedicated DVB-H networks is defined in comparison with the handover in converged DVB-H/UMTS (or other uplink technology networks) networks. Regarding to the handover in the dedicated DVB-H networks, the terminal does not utilize the uplink technology (e.g. UMTS) to aid handover process. Instead, the handover is only a passive handover where the terminal performs the whole handover process itself.

6.2 Handover in DVB-H consists of three stages: handover measurement, handover decision-making and handover execution. The most important stage within handover regarding to terminal power consumption is handover decision-making stage. Because constantly making handover measurement in the handover decision-making stage consumes battery power, reducing handover measurement frequency in the handover decision-making stage will improve the terminal battery power efficiency.

6.3 As DVB-H terminals are mostly battery powered devices, the key idea to design a soft handover algorithm for DVB-H is to predict the handover moment in the handover process thus reducing the handover measurement frequency and saving battery power.

6.4 One handover decision-making algorithm, such as context aware handover algorithm or location aided handover algorithm, has the limitation that the algorithm may work well in one specific environment but not all environments. In order to cope with different scenarios, the hybrid handover decision-making algorithm was designed to suite to different situations. The basic idea of the hybrid handover decision-making algorithm is that a central management module manages different algorithm modules in the terminal. When the terminal is located in a specific scenario, the corresponding module to that specific scenario is chosen. In this way, the handover decision-making algorithm is always the optimized algorithm. Such a hybrid algorithm would need a large data base of predefined decision-making algorithms and pre-training and test of the different handover decision-making algorithms are necessary.

Problems of Chapter 7

7.1 Because the SNR is calculated from the RSSI and the noise characteristics, it provides a more accurate estimation of the received effective signal than the RSSI only measurement.

7.2 A Cumulative Distribution Function (CDF) describes a statistical distribution. It gives at each possible outcome the probability of receiving that outcome or a lower valued one. Because the CDF gives a probability value, it depends not only on the current SNR but also on the SNR history of the signal. This not only eliminates the frequent handover phenomenon (such as the

Ping Pong effect) seen in instantaneous RSSI value based handover but also avoids the "fake signals" caused by frequency confusion. The "fake signals" can be eliminated because the "fake signals" are only caused by short period signals and evaluating the signals' history reduces the chances of making decisions on short period signals.

Problems of Chapter 8

8.1 Repeaters provide an efficient solution to increase the coverage of the broadcasting networks. In broadcasting networks, the network operators usually firstly put high power transmitters at strategic points to quickly ensure an attractive coverage and then, at a second step, increase their coverage by placing low power repeaters in the dead spots or shadow areas, such as tunnels, valleys, or indoor areas. A repeater is simply a device that receives an analogue or digital signal and regenerates the signal along the next leg of the medium.

8.2 There are two different kinds of repeaters. Passive repeaters and active repeaters. A passive repeater receives and re-transmits a DVB-H signal without changing the signalling information bits. The signal is only boosted. An active repeater can demodulate the incoming signal, perform error recovery and then remodulate the bit stream. The output of the error recovery can even be connected to a local remultiplexer to enable insertion of local programs. It means that the entire signal is regenerated. The repeaters used in RA_handover approach are active repeaters.

8.3 The main idea of RA_handover algorithm is to use intelligent active repeaters to aid the handover process in DVB-H. In the RA_handover algorithm, the active repeaters are located in the cell border area. Each active repeaters will broadcast with unique identification information to the terminals that move to the border area. In this way, the terminals do not need to make handover measurement except when they have received the active repeater information thus saving battery power consumption.

8.4 The possible drawback of the RA_handover algorithm is that the RA_handover algorithm need many active repeaters which may add additional cost to the network. However, considering that the main transmitter power can be reduced after the introduction of the active repeaters at the border areas and the increased quality of service brought by the active repeaters, the overall cost of the network maybe balanced. A detailed study of the cost comparison need to be done to evaluate the drawbacks of the RA_handover algorithm.

Problems of Chapter 9

9.1 The soft handover probability is calculated by taking the ratio of the area where the soft handovers are possible to the total network coverage area.

Problems of Chapter 10

10.1 In the converged DVB-H/UMTS networks, the DVB-H is used as a downlink technology to provide high bit rate IP data services while the UMTS is used as an uplink technology to provide interactive features. On the other hand, the converged DVB-H/UMTS networks can be used to address the congestion problems due to the limited multimedia channel access in the UMTS networks.

10.2 DVB-H will broadcast the same data services to as many users as possible within its coverage area due to its broadcast nature. When only a few users request the same service, using DVB-H as a downlink channel will waste bandwidth that could be utilized by many more users. In this case, UMTS is an efficient way to provide the service to the limited number of users. On the other hand, when the number of users requiring the same service exceeds a certain limit, DVB-H will be a profitable way to provide the downlink channel. The intersystem handover in the converged DVB-H/UMTS networks is defined as a radio resource relocation algorithm. It is used to allocate the optimum radio resources between DVB-H and UMTS respect to reducing the network operation cost while providing the required quality of service.

10.3 DVB-H is not only a broadcast technology. DVB-H targets mainly IP data services, called IP Datacast. It is assumed that a large number of the IP Datacast services will be of interactive nature. The main application area of DVB-H should be the so called converged networks where DVB-H is used as a downlink technology while another technology (e.g. UMTS) is used as an uplink.

10.4 DVB-H uses a broadcast way to deliver large quantities of popular content to large user groups at relatively low cost but is very limited in providing interactivity and personalization of content. On the other hand, UMTS does not provide a low cost delivery mechanism for large quantities of data simultaneously to large user groups but has the excellent properties of interactivity and personalization. Thus UMTS can complement DVB-H by providing a terrestrial interaction channel for DVB-H.

10.5 The converged DVB-H/UMTS is a multi-tier network. The cell structure of the converged DVB-H/UMTS is a hierarchical cell structure in which the DVB-H cells overlay the UMTS cells. In the converged DVB-H/UMTS networks, the DVB-H cells cover the same area as the UMTS cells do. Because the DVB-H cell size is usually larger than the UMTS cell size, one DVB-H cell normally covers several UMTS cells. Hence, the converged network has a two-layer macro-micro cell structure, in which the DVB-H cell is in the macro layer while the UMTS cell is in the micro layer.

10.6 In order for the terminal to be able to perform intersystem handover in the converged DVB-H/UMTS networks, either the terminal has two independent receivers or the terminal has a single dual mode receiver. For handheld devices, it will reduce complexity and cost to have a single dual mode receiver rather than have two independent receivers.

10.7 In the converged DVB-H/UMTS networks, the terminal utilizes the "compressed mode" of the UMTS to perform DVB-H signal strength measurement while utilizes the "time slicing mode" of DVB-H to perform UMTS signal strength measurement.

10.8 In its simplest and most useful form, a stochastic tree is a transition diagram for a continuous-time Markov chain, unfolded into a tree structure. It has been used as modelling tools to analyze medical treatment decisions for almost a decade.

10.9 The stochastic tree algorithm is used in the converged DVB-H/UMTS networks to deduct the network operation cost $COST$ and the network quality of service $QAST$. By trading off $COST$ and $QAST$, the intersystem soft handover decision between DVB-H and UMTS can be made.

Problems of Chapter 11

11.1 A hybrid broadcast network is a network where different broadcast standards are simultaneously used. The hybrid broadcast network makes the vertical handover between different broadcast standards become necessary.

11.2 The main mobile broadcast standards currently available around the world are: DVB-H (developed in Europe and deployed around the world), T-DMB (developed in Korea and deployed around the world), DMB-T (developed in China), ISDB-T (developed in Japan and deployed in Japan and Brazil), DVB-SH (developed in Europe), CMMB (developed in China).

11.3 The main idea of the vertical handover algorithm in the hybrid broadcast networks is to design a handover control layer between the IP layer and the lower Mac/Physical layer. As the upper layers above the handover control layer are almost the same for different broadcast standards. The main task will be how to detect and convert the signals from the lower layers of the different broadcast networks and to convert them into the IP packets of the common upper layers.

Problems of Chapter 12

12.1 As DVB-H technology is evolving, new research topics are coming up. Some of the potential research issues in DVB-H are: Cross layer design; DVB-H indoor reception; handover design; mobile client design.

References

1. J. Henriksson, P. Talmola (2004) Coach Potato. IEE Communications Engineer, Volume 2, Issue 4, pages 28-32
2. S. O'Leary, D. Priestly, N. McSparron (1998) Mobile Broadcasting of DVB-T Signals. Broadcasting, IEEE Transactions on, Volume 44, Issue 3, pages 346-352
3. I. Poole (2005) What Exactly Are ... DMB & DVB-H?. Communications Engineer, Volume 3, Issue 3, pages 44-45
4. Gwangsoon Lee, Sammo Cho, Kyu-Tae Yang, Young Kwon Hahm, Soo In Lee (1991) Development of Terrestrial DMB Transmission System Based on Eureka-147 DAB System. Consumer Electronics, IEEE Transactions on, Volume 51, Issue 1, pages 63-68
5. D. Lenton (2003) The Small Screen. IEE Review, Volume 49,Issue 10, pages 38-41
6. S. Hartwig, M. Luck, J. Aaltonen, R. Serafat, W. Theimer (2000) Mobile Multimedia - Challenges and Opportunities. Consumer Electronics, IEEE Transactions on, Volume 46, Issue 4, pages 1167-1178
7. J. Henriksson (2003) DVB-X. DVB-Scene Magazine, pages 7
8. G. Gardikis, A. Kourtis, P. Constantinou (2003) Dynamic Bandwidth Allocation in DVB-T Networks Providing IP Services. Broadcasting, IEEE Transactions on, Volume 49, Issue 3, pages 314-318
9. Digital Video Broadcasting Project (2005) TV in Hand. DVB-Scene Magazine, pages 2-15
10. A. B. Kempe (1879) On the Geographical Problem of the Four Colors. Amer. J. Math. pages 193-200
11. D. Gesbert, M. Shafi, Da-shan Shiu, P. J. Smith, A. Naguib (2003) From Theory to Practice: An Overview of MIMO Space-time Coded Wireless Systems. Selected Areas in Communications, IEEE Journal on, Volume 21, Issue 3 pages 281-302
12. Junqiang Li, K. B. Letaief, Zhigang Cao (2002) Adaptive Cochannel Interference Cancellation in Space-time Coded Communication Systems. Communications, IEEE Transactions on, Volume 50, Issue 10, pages 1580-1583
13. C. L. Miller, D. P. Taylor, P. T. Gough (2001) Estimation of Co-channel Signals With Linear Complexity. Communications, IEEE Transactions on, Volume 49, Issue 11, pages 1997-2005

14. Jun Wang, Zhixing Yang, Changyong Pan, Jian Song, Li Yang (2005) Iterative Padding Subtraction of the PN Sequence for the TDS-OFDM over Broadcast Channels, IEEE Trans. on Broadcasting, Vol. 51, No. 4. pages 1148-1152
15. Ming Kang, M. S. Alouini (2004) A Comparative Study on the Performance of MIMO MRC Systems With and Without Cochannel Interference. Communications, IEEE Transactions on, Volume 52, Issue 8, pages 1417-1425
16. Xavier Lagrange (1997) Multitier Cell Design. Communications Magazine, IEEE, Volume 35, Issue 8 pages 60-64
17. Tak-Shing, Peter Yum, Wing-Shing Wong (1993) Hot-Spot Traffic Relief in Cellular Systems. Selected Areas in Communications, IEEE Journal on, Volume 11, pages 934-940
18. Hongyi Wu, Chunming Qiao (2003) Modeling iCAR Via Multi-dimensional Markov Chains. Mobile Networks and Applications, Volume 8, Issue 3, pages 295-306
19. Gordon Hazen (2000) Preference factoring for stochastic trees. Management Science, Volume 46, Issue 3, pages 389-403
20. Gordon Hazen (1993) Factored Stochastic Trees: A Tool for Solving Complex Temporal Medical Decision Models, Medical Decision Making, 13, Pages: 227-236
21. G. B. Hazen, J. M. Pellissier (1996) Recursive Utility for Stochastic Trees. Operations Research, Volume 44, Issue 5, pages 788-809
22. Zi-Wei Zheng, Zhi-Xing Yang, Chang-Yong Pan, Yi-Sheng Zhu (2003) Cutoff Rate and Outage Probability Performance Comparisons Between DVB-T and DMB-T Systems Under Mobile Multipath Channels. Broadcasting, IEEE Transactions on, Volume 49, Issue 4, pages 390-397
23. T. Nicolay, M. Buchholz, M. Gross (2005) Baseband Signal Processing of Digital Video Broadcasting Direct-conversion Zero-IF Tuners. Consumer Electronics, IEEE Transactions on, Volume 51, Issue 1, pages 48-53
24. C. Skianis, G. Kormentzas, K. Kontovasilis (2005) Interactions Between Intelligent Multimodal Terminals and a Network Management System in a B3G Context, Special Issue on "WLAN/3G Integration for Next-Generation Heterogeneous Networks", Wireless Communications & Mobile Computing, Wiley
25. X. D. Yang, Y. H. Song, T. J. Owens, J. Cosmas, T. Itagaki (2006) Incorporating Digital Repeaters into the Soft Handover Standard in Digital Video Broadcasting - Handheld. the International Journal of Services and Standards
26. C. Drane, M. Macnaughtan, C. Scott (1998) Positioning GSM Telephones. IEEE Communications Magazine, pages 46-54
27. Xaodong Yang, T. J. Owens (2008) Intersystem Soft Handover for Converged DVB-H and UMTS Network. IEEE Transaction on Vehicular Technology
28. E. Telatar (1999) Capacity of Multi-antenna Gaussian Channels. European Trans. On Telecomm. ETT, Volume 10, Issue 6, pages 585-596
29. Young B. Choi, Joshua S. Krause, Michelle Arcibal Imperio, Stephen P. Macchio, Kimberly A. Rill (2005) Applications of "Human Factors" in Wireless Telecommunications Service Delivery. International Journal of Services and Standards, Volume 1, pages 287-298
30. S. Kanthi Herath, Atul Gupta (2005) A Framework for Analysing Cost Structures in Business Process Reengineering (BPR). International Journal of Services and Standards, Volume 1, pages 494-511

31. U. Reimers (1997) DVB-T:the COFDM-based System for Terrestrial Television. Electronics & Communication Engineering Journal, Volume 9, Issue 1, pages 28-32
32. Michael Speth, Stefan Fechtel, Gunnar Fock, Heinrich Meyr (2001) Optimum Receiver Design for OFDM-Based Broadband Transmission-Part II: A Case Study. IEEE Transactions on Communications, Volume 49, Issue 4, pages 571-578
33. A. Centonza, G.A. Taylor, T. Itagaki, T.J. Owens, J. Cosmas, Y.H. Song (2006) Stochastic User Behaviour Modelling and Network Simulation for Resource Management in Cooperation with Mobile Telecommunications and Broadcast Networks, International Journal of Mobile Communications (IJMC) Volume 4, Issue 1, Pages: 97-117
34. A.K. Salkintzis, C. Fors, R. Pazhyannur (2002) WLAN-GPRS Integration for Next-generation Mobile Data Networks, Wireless Communications, IEEE, Volume 9, Issue 5, Pages: 112-124
35. D. Kouis, D. Loukatos, K. Kontovasilis, G. Kormenztas, C. Skianis (2005) On the Effectiveness of DVB-T for the Support of IP-based Services in Heterogeneous Wireless Networks, Special Issue on Internet over Digital Broadcast Video Networks, Elsevier Science
36. Melanie Tory, Torsten Mller (2004) Human Factors in Visualization Research. IEEE Transactions on Visualization and Computer Graphics, Volume 10, Issue 1, pages 72-84
37. Antonio Servetti, Juan Carlos De Martin (2002) Perception-Based Partial Encryption of Compressed Speech. IEEE Transactions on Speech and Audio Processing, Volume 10, Issue 8, pages 637-643
38. Pollini, G.P. (1996) Trends in Handover Design. Communications Magazine, IEEE, Volume 34, Issue 3, Pages 82 - 90
39. N. P. Stijn (2003) Study of Soft Handover in UMTS. Master Thesis, Danmarks Tekniske Universitet, Danmark
40. Andres Arjona (2005) Internet Protocol Datacasting - Transparent Interactivity Using Different Communication Channels. Master Thesis, Helsinki University of Technology, Helsinki, Finland
41. Yue Chen (2003) Soft Handover Issues in Radio Resource Management for 3G WCDMA Networks. PhD Thesis, University of London (Queen Mary)
42. Arto Hamara (2005) Considerations on Loss-Free Handover in IPDC over DVB-H Networks. Master Thesis, University of Turku, Finland
43. Lawrence Harte (2004) Introduction to GSM: Physical Channels, Logical Channels, Network, and Operation. ALTHOS Publishing
44. J. Laiho, A. Wacker, T. Novosad (2002) Radio Network Planning and Optimisation for UMTS. JOHN WILEY & SONS, LTD
45. U. Reimers (2004) Digital Video Broadcasting, 2nd edition. Springer, Berlin Heidelberg New York
46. Harri Holma, Antti Toskala (2001) WCDMA for UMTS Radio Access for Third Generation Mobile Communications. John Wiley & Sons
47. Janevski, Toni (2003) Traffic Analysis and Design of Wireless IP Networks. Artech House
48. C. Drane (1998) Positioning Systems in Intelligent Transportation Systems. Artech Hause, Boston, London

49. William H. Press, Brian P. Flannery, Saul A. Teukolsky, William T. Vetterling (1992) Numerical Recepes in C: The Art of Scientific Computing, Cambridge University Press, pages 650-655
50. DAB Organization (2007) http://www.worlddab.org
51. DVB Project Office (2007) http://www.dvb.org
52. UMTS-forum (2007) http://www.umts-forum.org
53. IPDC Forum (2007) http://www.ipdc-forum.org
54. Celtic Organization (2005) http://www.celtic-initiative.org
55. Finish Mobile TV Organization (2005) http://www.finishmobiletv.com
56. Dutch Mobile TV Organization (2005) http://www.dutchmobiletv.nl
57. OPNET (2007) http://www.opnet.com
58. Nokia (2007) http://www.nokia.com/mobiletv
59. UDCAST (2005) http://www.udcast.com
60. EBU (2005) http://www.ebu.ch/en/technical/trev/trev_home.html
61. IST CISMUNDUS Project (2005) http://dea.brunel.ac.uk/project/Cismundus
62. DTG (2005) http://www.dtg.org.uk
63. Austria DVB-H Pilot Project (2007) http://www.mobiletvaustria.at
64. IETF RFC 2327 (1998) SDP: Session Description Protocol
65. 3GPP (2005) 3rd Generation Partnership Project; Technical Specification Group Services and System Aspects; Multimedia Broadcast/Multicast Service (MBMS); Architecture and Functional Description (Release 6). V6.7.0, TS 23.246
66. ETSI (2005) Digital Cellular Telecommunications Systems (Phase 2+); Universal Mobile Telecommunications System (UMTS); Handover Procedures. V6.1.0
67. ETSI (2004) Digital Video Broadcasting (DVB): Transmission System for Handheld Terminals (DVB-H) V1.1.1, EN 302 304
68. ETSI (2005) Digital Video Broadcasting (DVB): DVB-H Implementation Guidelines V1.2.1, TR 102 377
69. ETSI (2004) Digital Video Broadcasting (DVB): Guidelines on Implementation and Usage of Service Information (SI), V1.6.1, TR 101 211
70. ETSI (2004) Digital Video Broadcasting (DVB): framing structure, channel coding and modulation for digital terrestrial television. V1.5.1, EN 300 744
71. ISO/IEC (2000) Information Technology - Generic Coding of Moving Pictures and Associated Audio Information: Systems, 13818-1
72. ETSI (2007) Digital Video Broadcasting (DVB): Specification for Service Information (SI) in DVB Systems, V1.8.1, EN 300 468
73. ETSI (2005) Digital Video Broadcasting (DVB): Transmission to Handheld Terminals (DVB-H); Validation Task Force Report. V1.1.1, TR 102 401
74. ETSI (2004) Digital Video Broadcasting (DVB): DVB Specification for Data Broadcasting. V1.4.1, EN 301 192
75. ETSI (2005) Digital Audio Broadcasting (DAB); DMB Video Service; User Application Specification. V1.1.1, TS 102 428
76. ETSI (2004) Digital Video Broadcasting (DVB): Implementation Guidelines for DVB Terrestrial Services; Transmission Aspects. TR 101 190, V1.2.1
77. ETSI (2006) Digital Video Broadcasting (DVB): IP Datacast over DVB-H: Electronic Service Guide (ESG). TS 102 471, V1.2.1
78. OMA BCAST (2007) OMA-TS-BCAST_Service_Guide-V1: Service Guide for Mobile Broadcast Services
79. OMA BCAST (2007) OMA-TS-BCAST_DVB_Adaptation-V1: Broadcast Distribution System Adaptation - IPDC over DVB-H

80. OMA BCAST (2007) OMA-TS-BCAST_MBMS_Adaptation-V1: Broadcast Distribution System Adaptation - 3GPP/MBMS Specification
81. OMA BCAST (2007) OMA-TS-BCAST_BCMCS_Adaptation-V1: Broadcast Distribution System Adaptation - 3GPP2/BCMCS Specification
82. OMA BCAST (2007) OMA-TS-BCAST_Services-V1: Mobile Broadcast Services Specification
83. OMA BCAST (2007) OMA-TS-BCAST_Distribution-V1: File and Stream Distribution for Mobile Broadcast Services Specification
84. 3GPP (2004) Multiplexing and Channel Coding (FDD). V6.2.0, TS25.212, pages 55-58
85. ITU (2001-2003) Method for Point-to-area Predictions for Terrestrial Services in the Frequency Range 30 MHz to 3000 MHz. ITU Recommendation ITU-R P.1546-1
86. ETSI (2005) Digital Cellular Telecommunications System (Phase 2+); Multiplexing and Multiple Access on the Radio Path (3GPP TS 45.002 Version 6.12.0 Release 6). V6.12.0, ETSI TS 145 002
87. ETSI (2006) Digital Video Broadcasting (DVB): IP Datacast over DVB-H: Program Specific Information (PSI)/Service Information (SI). TS 102 470, V1.1.1
88. DVB Document A100 (2005) IP Datacast over DVB-H: Service Purchase and Protection (SPP)
89. DVB Document A111 Rev.1 (2007) Framing Structure, Channel Coding and Modulation for Satellite Services to Handheld Devices (SH) below 3 GHz
90. DVB Document A110 Rev.1 (2007) System Specifications for Satellite Services to Handheld Devices (SH) Below 3 GHz
91. DVB Document A113 (2007) IPDC over DVB-H: Content Delivery Protocols (CDP) Implementation Guidelines
92. DVB Document A112 (2007) IPDC over DVB-H: Electronic Service Guide (ESG) Implementation Guidelines
93. DVB Document A096 (2005) IPDC over DVB-H: Set of Specifications for Phase 1
94. G. May (2004) The IP Datacast System - Overview and Mobility Aspects. Consumer Electronics, 2004 IEEE International Symposium on, pages 509-514
95. U. Reimers (1998) Digital Video Broadcasting. Communications Magazine, IEEE Volume 36, Issue 6 pages 104-110
96. Carlos Herrero, Petri Vuorimaa (2004) Delivery of Digital Television to Handheld Devices Proceedings of the 1st International Symposium on Wireless Communication Systems, ISWCS 2004 pages 240-244. Port Louis, Mauritius
97. E. Launay, J. Hammerschmidt, J. Rinne (1999) Antenna Diversity Techniques for Mobile and Portable DVB-T Receivers. Proceedings of the International Broadcasting, pages 417-422
98. Mike Brooks (2005) Development of Broadcast Technologies for Mobile TV. The IEE Seminar on Broadcasting Spectrum - the Issues, London,UK
99. I. Gaspard (1999) Mobile Reception of the Terrestrial DVB-system. Vehicular Technology Conference,1999 IEEE 49th Volume 1, pages 151-155
100. Aurelian Bria, David-Gomez Barquero (2005) Scalability of DVB-H Deployment on Existing Wireless Infrastructure. Proceedings of the IEEE PIMRC
101. X. D. Yang, Y. H. Song, T. J. Owens, J. Cosmas, T. Itagaki (2004) Performance Analysis of Time Slicing in DVB-H. Sympotic2004, pages 183-186
102. Gunther May (2005) Loss-free Handover for IP Datacast over DVB-H Networks. IEEE ISCE 2005, Macau

103. K. Mitchell, K. Sohraby (2001) An Analysis of the Effects of Mobility on Bandwidth Allocation Strategies in Multi-Class Cellular Wireless Networks, INFOCOM 2001, Proceedings, Volume 2, Page(s):1005-1011
104. Jani Väre, Matti Puputti (2004) Soft Handover in Terrestrial Broadcast Networks. MDM2004. Proceedings, pages 236-242
105. L. Schwoerer, J. Vesma (2003) Fast Scattered Pilot Synchronization for DVB-T and DVB-H. Proc. 8th International OFDM Workshop, Hamburg, Germany
106. X. D. Yang, Y. H. Song, T. J. Owens, J. Cosmas, T. Itagaki (2004) Seamless Soft Handover in DVB-H Networks. Softcom2004, Proceedings, pages 27-30
107. X. D. Yang, Y. H. Song, T. J. Owens, J. Cosmas, T. Itagaki (2004) Performance Analysis of OFDM Scheme in DVB-T. Emerging Technologies: Frontiers of Mobile and Wireless Communication, 2004, Proceedings of the IEEE 6th Circuits and Systems Symposium on, volume 2, pages 489-492
108. M. Kornfeld (2004) DVB-H - the Emerging Standard for Mobile Data Communication. Consumer Electronics, 2004 IEEE International Symposium on, pages 193-198
109. Jean-Luc Sicre, Erwan Launay (2005) Is IP - Multicasting Over DVB-H a Real Business Enabler for a Mobile Telco Operator? Multiradio Multimedia Communications 2005,
110. J. Aaltonen, H. Pekonen, T. Auranen, K. Laiho, P. Talmola (2002) Power Saving Considerations in Mobile Datacasting Terminals. Proceedings of IEEE International Symposium on Consumer Electronics, pages 43-48
111. Jani Vare, Arto Hamara, Jarno Kallio (2005) Approach for Improving Receiver Performance in Loss-free Handovers in DVB-H Networks. Globecom2004, Volume 5, pages 3326-3331
112. X. D. Yang, Y. H. Song, T. J. Owens, J. Cosmas, T. Itagaki (2005) An Investigation and a Proposal for Handover Decision-making in DVB-H. 14th IST Mobile and Wireless Communications Summit, Dresden, Germany
113. L. Schwoerer (2004) Fast Pilot Synchronization Schemes for DVB-H. Proceedings of the 4th IASTED International Multi-Conference on Wireless and Optical Communications, Banff, Canada
114. M. Dohler, A. D. Marbini, H. Aghvami (2001) Intelligent base station hand-over algorithm based upon a precise power drop model. Proceedings of VTC2001 Spring
115. Aurelian Bria, David-Gomez Barquero (2005) Scalability of DVB-H Deployment on Existing Wireless Infrastructure. IEEE PIMRC
116. C. Trolet (2002) SPOT: Filling Gaps in DVB-T Networks With Digital Repeaters. BroadcastAsia2002 International Conference
117. M. Schinnenburg, I. Forkel, B. Haverkamp (2003) Realization and Optimization of Soft and Softer Handover in UMTS Networks. Personal Mobile Communications Conference, 2003 5th European (Conf. Publ. No. 492) pages 603-607
118. M. Benson, H. J. Thomas (2002) Investigation of the UMTS to GSM Handover Procedure. Vehicular Technology Conference, VTC Spring 2002 IEEE 55th, pages 1829-1833
119. Delphine Lugara, Lionel Ammi, Marc Griguer, J. Tartière (2004) UMTS to GSM Handover Based on Compressed Mode Technique. Communications 2004, IEEE International Conference on, Volume 5, pages 3051-3055
120. V. K. Varma, S. Ramesh, K. D. Wong, M. Barton, G. Hayward, J. A. Friedhoffer (2003) Mobility Management in Integrated UMTS/WLAN Networks.

Communications 2003, IEEE International Conference on, Volume 2, pages 1048-1053
121. M. Jaseemuddin (2003) An Architecture for Integrating UMTS and 802.11 WLAN Networks. Computers and Communication 2003, Proceedings. Eighth IEEE International Symposium on, Volume 2, pages 716-723
122. R. Burow, K. Fazel, P. Hoeher, O. Klank, H. Kussmann, P. Pogrzeba, P. Robertson, M. J. Ruf (1998) On the Performance of the DVB-T System in Mobile Environments. Global Telecommunications Conference, 1998. GLOBE-COM - 98. The Bridge to Global Integration, IEEE, Volume 4, pages 2198-2204
123. J. T. Ong, S. V. B. Rao, G. Shanmugam, Yan Hong (2004) Spatial Variation of Mobile DTV Signals in Singapore. RF and Microwave Conference, RFM 2004, pages 28-32
124. J. Tavares, A. Navarro (2005) Optimal IP Packet Length for DVB-T Transmission. IEEE ISCE 2005, Macau
125. S. Mangold, S. Kyriazakos (1999) Applying Pattern Recognition Techniques Based on Hidden Markov Models for Vehicular Position Location in Cellular Networks. Vehicular Technology Conference, VTC 1999 - Fall, IEEE VTS 50th, Volume 2, pages 780-784
126. Bob Tyler (2005) New Demands of Broadcasting for Spectrum. The IEE seminar on broadcasting spectrum the issues, London
127. Sang Yoon Park, Nam Lk Cho, Sang Uk Lee, Kichul Kim, Jisung Oh (2001) Design of 2K/4K/8K-point FFT Processor Based on CORDIC Algorithm in OFDM Receiver. Communications, Computers and Signal Processing, 2001. PACRIM. 2001 IEEE Pacific Rim Conference on, Volume 2, pages 457-460
128. Peter Unger, Thomas Krner (2005) Radio Network Planning of DVB-H/UMTS Hybrid Mobile Communication Networks. 14th IST Mobile and Wireless Communications Summit, Dresden, Germany
129. H.Schulzrinne, S.Casner, R.Frederick, V.Jacobson (2003) RTP: A Transport Protocol for Real-Time Applications. IETF RFC 3550
130. T.Paila, M.Luby, R.Lehtonen, V.Roca, R.Walsh (2004) FLUTE - File Delivery over Unidirectional Transport. IETF RFC 3926
131. Hanli Zou, B. McNair, B. Daneshrad (2001) An Integrated OFDM Receiver for High-speed Mobile Data Communications. Global Telecommunications Conference, 2001. GLOBECOM'01, IEEE, Vol.5, Pages 3090-3094
132. P.J. Czerepinski, T.M. Chapman, J. Krause (2004) Coverage and Planning Aspects of MBMS in UTRAN. Fifth IEE International Conference on 3G Mobile Communication Technologies, Pages 529-533
133. DVB Project (2004) TM3095R2:Technical Requirements for IP Datacast. htpp://www.dvb.org
134. DigiTAG (2005) Television on a Handheld Receiver V1.1. http://www. digitag.org
135. Medialab (2005) http://www.medialab.sonera.fi/workspace/MBMSWhite Chapter.pdf
136. Peter Pogrzeba, Ralf Burow, Gerard Faria, Andrew Oliphant (2005) Lab & Field Tests of Mobile Applications of DVB-T, http://www.harris.com
137. IST Integrated Project INSTINCT (2005) http://www.ist-instinct.org
138. BMCO Forum (2007) http://www.bmcoforum.org
139. DVB-H (2007) http://www.dvb-h.org/services.htm
140. IST STREP Project MING-T (2007) http://www.ming-t.org
141. CMMB Group (2007) http://www.cmmb.org.cn

142. WING-TV Project (2005) http://www.celtic-initiative.org/Projects/WING-TV/default.asp
143. K. Matsumura, K. Sakaki, K. Otsuki, H. Kokubun, K. Fujisawa (2005) A Practical Experiment on Mobile Data Broadcasting. IEEE Second Consumer Communications and Networking Conference, CCNC pages 337-342
144. Digitalradiotech (2005) http://www.digitalradiotech.co.uk/dvb-h_dab_dmb.htm.
145. Andreas Sieber, Chris Weck (2004) What's the Difference Between DVB-H and DAB - in the Mobile Environment? EBU, Technical Review
146. Gerard Faria (2004) DVB-H to Deliver Digital TV to Hand-held Terminals. http://www.broadcastchapters.com/BcastAsia04/BAsia04TeamcastDVBH.pdf
147. Michael Kornfeld, Ulrich Reimers (2005) DVB-H - the Emerging Standard for Mobile Data Communication. EBU, Technical Review
148. Nokia (2005) http://press.nokia.com/PR/200509/1010115_5.html
149. Qualcomm (2005) http://www.qualcomm.com/mediaflo/index.shtml
150. Toivo Kivirinta, Timo Ali-Vehmas, Teemu Mutanen, Tero Tuominen, Matti Vuorinen (2004) Forecasting Market Demand For Mobile Broadcast Services in Finland. Helsinki University of Technology, Final Report, Operations Research Project Work Seminar
151. George Hufford (1968) The ITS Irregular Terrain Model, Version 1.2.2 The Algorithm. NTIA/ITS
152. Digital Video Broadcasting Project (2003) Commercial Requirements: IP Datacast in DVB-H. TM2967 (SB1299)
153. The Australian Broadcasting Authority Benjamin Offices (2005) Digital Terrestrial Television Broadcasting Planning Handbook - Including Technical and General Assumptions. Second Edition
154. Caj Sodergard (2003) Mobile Television - Technology and User Experiences: Report on the Mobile - TV Project. VTT Technical Research Centre of Finland
155. Driving Standards Agency (2004) The Highway Code. The Stationery Office Books
156. (2005) http://www.cell-alert.co.uk/cell_broadcasting.htm
157. Philips (2000) TDA10045H DVB-T Channel Receiver. Philips Semiconductors Product specification
158. J. Henriksson (2003) DVB-H Outline. Slideset prepared by Dr. Jukka Henriksson, available from http://www.dvb.org
159. EBU (2005) EBU Guidelines for the RRC-06. EBU Technical Information I37-2005
160. Gerard Faria (2005) DVB-H: Digital TV in The Hands. http://www.dvb-h-online.org/technology.htm
161. DVB-H Group (2005) http://www.dvb-h-online.org
162. NHK (20007) http://www.nhk.or.jp
163. T-Systems (2005) Parameters of DVB-H Transmission T-Systems Media&Broadcast. http://www.dvb-h-online.org
164. H. Joki, J. Poikonen (2005) Analysis and Simulation of DVB-H Link Layer. TUCS Technical Report, No. 695
165. MediaFLO (2007) MediaFLO FLO Technology Overview. Qualcomm

Index

MIX
Papier aus verantwortungsvollen Quellen
Paper from responsible sources
FSC® C105338

If you have any concerns about our products,
you can contact us on
ProductSafety@springernature.com

In case Publisher is established outside the EU,
the EU authorized representative is:
Springer Nature Customer Service Center GmbH
Europaplatz 3, 69115 Heidelberg, Germany

Printed by Libri Plureos GmbH
in Hamburg, Germany